RHYMES TO PREDICT THE WEATHER

Rhymes to Predict the Weather

By Don Haggerty

Springmeadow Publishers · Seattle

Copyright © 1985 by Don Haggerty

All rights reserved. No part of this publication may be reproduced, stored in a retrieval system, or transmitted, in any form or by any means, electronic, mechanical, photocopying, recording, or otherwise, without the prior written permission of the publisher.

Printed in the United States of America
Library of Congress Catalog Card Number: 84-52672
ISBN: 0-9614703-0-5

First Edition

Inquiries should be addressed to

Springmeadow Publishers
P.O. Box 31038
Seattle, Washington 98103

To the clouds,
whose endless dance
above my head
has earned my endless
applause.

Table of Contents

1. **Introduction to Weather and Forecasting Principles**   1
   - All the Sky is a Stage   1
   - The Basics of Weather Prediction   4
   - How to Use This Book   8

2. **Air Movement: How Weather Gets Its Start**   11
   - Warm Air Rises   11
   - Winds Flow From High to Low   14
   - The Way of Moisture   18
   - Other Lifting Forces   20

3. **Coriolis: The "Swirling" Force**   24

4. **Global Winds and General Air Circulation**   30
   - The Westerlies   30
   - Seasons   35
   - Air Masses   37

5. **Weather System Winds**   41
   - Surface Friction   41
   - Spiraling Winds and the Westerly Stream   44

6. Mentally Positioning Yourself Within a Weather System — 47
- The Wave Cyclone 47
- Warm Fronts 52
- Cold Fronts 56
- Identifying Frontal Passage 57
- Occluded Fronts 58
- Storm Quadrant Characteristics 59
- The Eddy Cyclone 62

7. Wind — 65
- Wind Basics 65
- Speed Indicators 67
- The Upper-Air Winds 69
- Winds: Fair and Foul 73
- Wind Shifts 74

8. Clouds — 78
- Identifying Clouds 78
- Cloud Tales 82
- Thunder Storms 86
- More Cloud Tales 87

9. Barometric Pressure — 91
- The Barometer in General 91
- Reading the Ups and Downs 92

10. Temperature — 96
- Stable and Unstable Air 96
- Unseasonable Temperatures 99
- Wind Chill 100

11. Humidity and the Dew Point   **102**
    Understanding Humidity, Dew, and the Dew Point.................... 102
    Estimating the Dew Point 104
    Tales of Dew, Fog, Thunder and Frost ... 107

12. Other Weather Phenomena   **112**
    Smell 112
    Sight and Sound 113
    Halos 114
    The Moon and Stars 115
    Sunsets 116
    Rainbows and Bees 118

INDEX   **121**

Chapter 1

Introduction to Weather and Forecasting Principles

All the Sky is a Stage

> The sky is a stage,
>   and high overhead,
> A drama unfolds
>   like a large table spread.
> With morsels so tasteful,
>   a feast for the eye,
> No stage can compare
>   to the matchless sky.

For a show so grand as the one offered us by the weather, you would think that by now someone would surely have conjured a way to charge admission to those wishing to partake of it. But, no. Admission is still free. Good seats are plentiful. There are no intermissions or commercial interruptions. And to top it all, not only does the weather still offer us the best show in town—it remains today the biggest show in the whole wide world!

And what sort of patrons might we expect to find among those whose names grace the guest book of this magnificent theatre? Stout peasants of snowy Himalayan peaks. Gritty merchants on the deserts of North Africa. Copper-clad natives of remote Pacific islands. Pomp royalty of European courts. Red-faced farmers of North American plains. Bare-bodied inhabitants of South American jungles. And pretty people of the world's higher societies, gazing from limousine sun roofs and penthouse gardens. Really, our names are all there.

> As a fish o'er whose head
>   surge the waves of the sea,
> In our great sea of air,
>   the fish becomes me.
> And high o'er my head
>   waves of air churn and roll;
> Deep currents, deep tides
>   move the air, move my soul.

We live in houses. We study in schools. We work in buildings. And we drive in cars. But where is it that we really do all this working and breathing and eating and sleeping and living?—In the great sea of air. It's our home, our ceiling, our walls, our windows.

All the Sky is a Stage

> At the close of the day,
> Pull your wits together:
> Put aside the TV,
> And watch the weather.

Of the puzzles to which man has put his mind to decipher, few there are that have left him indifferent, unmoved. But puzzles alone don't possess inherent attractions. The key is in the "putting of his mind" to the wonders before him—the challenge to capture meaning in apparently meaningless events. Then once grasped, the pleasure comes of understanding, of comprehending, the forces at work in his world, and the powers that move amidst the inner depths of his soul.

The weather offers such a puzzle. And comprehending a thing of such vastness and beauty as is contained within the limits of our atmosphere offers compensation in the fullest measure to any who would make efforts to acquaint oneself with its mysteries.

> Rain plays percussion
> As Wind sings a tune,
> Clouds dance to the music
> The whole afternoon.

To some, weather doesn't matter. It's a thing to be tolerated. Clouds are noted only when their rains threaten inconvenience, and winds observed only when their chill means discomfort.

But others see the weather through different eyes. To the farmer, weather means crops and livestock—or the destruction of both. To the seaman, weather means smooth and speedy sailing—or loss of time, revenue, and perhaps even life. To the airman, weather means safe, enjoyable flight—or long detours and closed airfields. To these people, weather matters. And that's the beginning of good forecasting.

> If today's sunny skies
> Bring you loud acclaim,
> Watch out lest tomorrow's skies
> Bring you the blame.

There is a shortcoming in trying to predict anything. That is, whatever you may predict, there's always the possibility that it won't happen. Simple as that, huh? Well, there's another side to the picture. Through your growing acquaintance with the weather and with the principles by which it operates, predictions you make will start to come true. You'll see that the goings-on of our atmospheric ocean really do have form and meaning. You'll see that, while the weather is wonderfully free to choose its course, there remain certain rules by which it must abide.

The Basics of Weather Prediction

The rules by which the weather sets its stage and costumes its actors are the subject of the rhymes contained in this book. But *applying* these rhymes involves several things:

1. The first step is the learning of certain basic weather principles. Chapters 2 through 6 make up the "why and wherefore" section of this book which will lead you through these principles and prepare you for

The Basics of Weather Prediction

the second, and primary, aspect of weather analysis and forecasting.

2. Observation. Seems obvious, doesn't it? But it is on this point that both the harsh and the gentle realities of weather appreciation come into focus. The harsh reality is that a knowledge of what makes the atmosphere tick can't come with just a few quick glances at the sky. Only time and experience can supply such know-how. But the gentle reality is this: By taking time to *observe* the weather and its ways, one comes to understand much more than just the physics of meteorological phenomena. A far greater reward awaits the one who, through quiet scrutiny and an unhurried spirit of appreciation, comes to find in the weather a special something on which to raise higher thoughts and somehow step for at least a time above the mayhem that has come to characterize our earth "down here on the ground."

As you can see, observation as we're using the term here does not mean isolated glances at the sky. Perhaps it would be clearer to say that, along with the principle of observation, there are the physical acts of "making observations" that are essential to good weather understanding. The plural is intentional. Making observations means observing the *trends* of the weather by connecting a series of observations into a meaningful flow of weather events. In its most basic form, one of these observations might mean:

- Glancing at the barometer, noting whether it has risen or fallen since your last check.

- Stepping outside to smell and feel the air, alert for changes in humidity and temperature.

- Testing the wind for direction and strength.

- Looking at the clouds long enough to judge their form, speed and direction.

This making of weather observations leads us then to the third aspect of our forecasting process.

3. Here we are ready to take the observations we have made and begin matching them up with what we know about the weather and its general behavior. Through this matching process, we can come up with a reasonable prediction of how the weather might develop. Chapters 7 through 12 are meant to enable just that. They are called the "if ... then" chapters, for it is here that your observations take on predictive action when, for instance, you ponder, "If the clouds appear thus, and the wind is blowing from there, then ..." Needless to say, once you have acquainted yourself with the "whys and wherefores" of weather behavior, the "if ... then" rhymes of this book are sure to be your most useful—and enjoyable—weather companions.

> If it happened once
> under these conditions,
> Well, the weather too
> follows certain traditions.

The principle here is one of historical sequence. And it's a principle that should be used to color all

The Basics of Weather Prediction 7

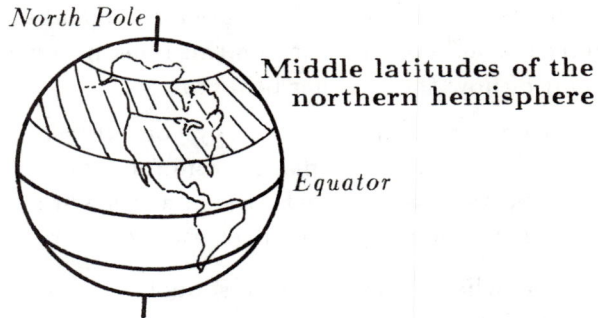

Figure 1.1: The atmospheric events discussed in this book describe weather as it generally occurs in the middle latitudes of the northern hemisphere.

your weather observations. Throughout this book you will find rhymes that touch on all the essentials of weather activity. You must be aware, however, that the weather principles as they are put forth here are a description of things as they *generally* happen here in the middle latitudes of the northern hemisphere. (See Figure 1.1). Local discrepancies *will* occur. Experience alone can tell you that, while such-and-such is the way the weather generally acts, thus-and-so is what happens in my back yard. Fine-tuning your senses to the peculiarities of your own local weather conditions is a pleasure I'm sure you will enjoy for the rest of your life.

Use the rhymes to gain a good understanding of how the weather generally acts, then temper that understanding with the real stuff of weather forecasting—

your own observations and experience. That's what the blank margins of this book are for—to make notes, to record observations, to turn this book into *your* account of how the weather behaves.

> One indicator makes lucky your guess.
> Two indications make errors much less.
> So take ye the weather sign at its word,
> If you look again and see a third.

The whole purpose of this book as a weather-watcher's field guide and reference manual is to supply you, first, with a means of learning just what the weather is and why it acts the way it does. Second, it will provide you with a concise but thorough "memory bank" of weather principles and forecasting rules that will be at your disposal whenever you may need or want them. The format of the book is designed to make all this as simple and enjoyable as possible.

How to Use This Book

Here's how this book can work for you. Start with the Table of Contents. There you'll find an outline of what the book has to offer. The items listed there are not arbitrary; rather, they progress in a logical order through the basic weather principles. Each of these principles is necessary for a firm understanding that will then enable application of the rhymes found in later chapters. Regular reference to the Table of Contents will help keep the information organized in your own mind as well as acting as a guidepost to locating rhymes on subjects of immediate interest.

How to Use This Book 9

The Table of Contents, however, is obviously lacking in thoroughness. That is why a thorough index is provided. A brief glance through the index at the back of the book will tell you immediately that it is not just an ordinary index. The terms listed there are not only the general weather terms used within the book. Just as the rhymes are provided as "pegs" on which to hang your memory of weather principles, so the index contains all the words that your memory may want to use to lead you back to the particular rhyme that deals with a particular principle. Thus, the word "soap," though it may have no apparent relation to weather, is included in the index as a pointer to a rhyme that makes reference to rising air that "squeezes upward like a wet bar of soap." So, if the word "soap" happens to be the key word by which you locate a certain rhyme dealing with the principle of rising air, the index will have served its purpose.

Once a certain rhyme on a subject has been located, others on the same subject can easily be found by noting key words of interest, then returning to the index and looking up those words. One rhyme will lead you to others, which will lead you to others, and so on. You'll find this use of the index especially handy because in a book of this nature there is no absolute method for grouping all the material on a particular topic. The chapter on temperature, for instance, is relatively short, but not because temperature is a small topic. Rather, temperature is such a vast subject and so interrelated with others that you'll find references to it scattered throughout the book. The particular placement of a rhyme in each instance is determined only by the overall flow of the book as it progressively lays out the rules of the weather and the principles by which to predict its activities.

The book you hold in your hand is written with you, the reader, in mind. While it is meant as a handy reference guide to weather principles and forecasting, it is also meant to provide you just plain enjoyable reading if you should choose to read it as books are normally read—starting from the front and reading to the end.

And, oh yes, bear in mind through all of your reading and sky watching, there's a whole lot more to the weather than just the rain drops that wet your roof and the sunshine that dampens your brow.

>   Florida frost
>   brings high orange cost.
>   South unseasonable
>   makes cotton unreasonable.
>   A wet spring for wheat,
>   the less bread we'll eat.
>   Dry summer for corn—
>   ham nor bacon each morn.

Well, so much for weather introductions. Enjoy yourself... and happy gazing!

Chapter 2

Air Movement: How Weather Gets Its Start

Warm Air Rises

> All winds are caused by just one thing—
> Differences in temperature arguing.

And that, in a nutshell, is quite literally the "high" and "low" of it, as you will soon see. If it wasn't for the on-going argument continually on the rise between the hot and cold spots of our earth, though the ol' globe might go on turning, our winds would come

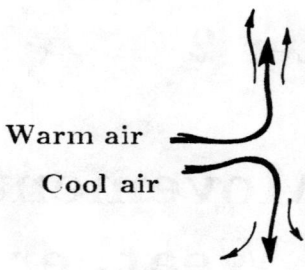

Figure 2.1: Warm air rises, cool air sinks.

to a stagnant halt. But fortunately for us the winds *haven't* stopped, and here is basically why:

> Warm air's light,
> Cool air's not.
> Warm takes flight,
> Cool walks a lot.

I'm sure you've seen ashes rise into the air above a camp fire. You've noticed smoke rising from a chimney. Steam rising from a teapot hasn't escaped your eye. And cold fog rolling from the base of the door on the stand-up freezer has spoken of more to you than just energy loss.

When we speak of the weather's starting place, the principle illustrated in Figure 2.1 is *it*. Here's what happens: As the world goes churning about in space, making its way around the sun each year, the sun shines continuously on the earth with its rays

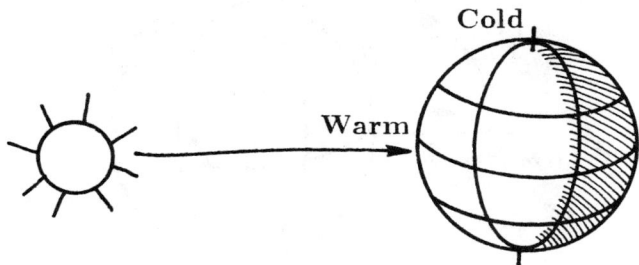

Figure 2.2: The earth's middle is warmed by the rays of the sun.

striking most directly at the equator (Figure 2.2). Due to all this continuous shining, things on that part of the earth's middle naturally tend to heat up a bit. The poles, on the other hand, are just the opposite. They're cold. In keeping with our principle, then, air at the equator has a tendency to always be rising (taking flight) while air at the poles is always sinking (walks a lot).

While warm air is busily rising at the equator, cold air is steadily making its way from the poles to replace the warm air that has gone aloft (Figure 2.3). It's almost as if the sun positions one end of a great big straw at the equator and then begins to inhale through it. Sure enough, warm air rises through the straw, but just as surely, cooler air from all around comes rushing in to where the warm air used to be.

Warm air rises. That's principle number one. But,

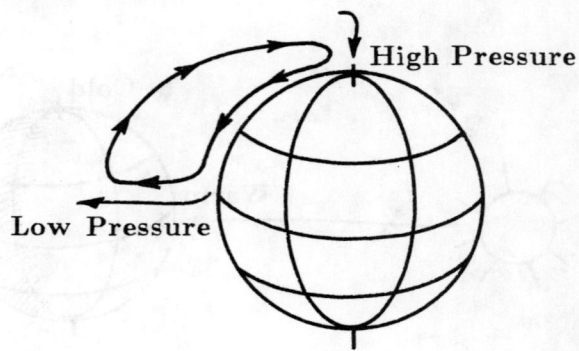

Figure 2.3: Warm air rises at the equator, while cold polar air flows to replace it. These temperature differences are what in turn bring about variations in atmospheric pressure.

as you can well see through the giant-straw example, differences in temperature which started the air to rising have now led to a difference in something else—a difference in atmospheric *pressure*. (See Figure 2.3).

Winds Flow From High to Low

> High-pressure air comes rolling low;
> Low-pressure air goes spinning high.
> The winds roll down, then up again,
> A fantasy ferris wheel in the sky.

The up and down motion of air is first brought about by differences in temperature causing air to rise where it is warm and to sink where it is cool

Winds Flow From High to Low

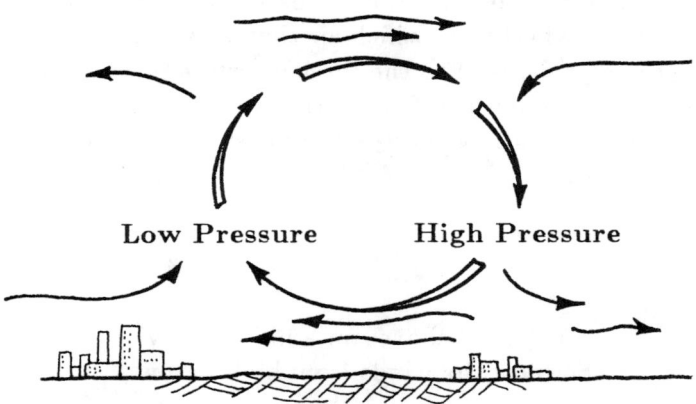

Figure 2.4: The great atmospheric ferris wheel demonstrates the stirring mechanism of the air.

(Figure 2.3). Where air rises, a partial vacuum is formed at the surface resulting in an area having "low atmospheric pressure." (Such an area of rising air is identified by a low reading on the barometer). Sinking air produces the opposite effect. Here, air tends to "pile up" as it descends toward the earth's surface, resulting in *high* atmospheric pressure. And when that's the case, the total weight of all that air piled overhead produces a *high* barometer reading.

The "down" side of the great atmospheric ferris wheel is known as the high-pressure side (Figure 2.4). Slowly sinking air found here is traveling on its way to fill the relative vacuum caused by a nearby low-pressure area. Air then begins to rise at the low's center, drawing air from all around its base to replenish what rises out its top. This is the "up" side of the ferris wheel from which air then flows outward to once again descend to the earth's surface in search of another trip upward. This is the great stirring mechanism of the atmosphere— the never-ending cycle of the winds.

> As a sled behaves on a hill of snow,
> So blow the winds from high to low.

When depicted on a weather chart, all these areas of falling and rising air—or areas of high and low pressure—take on an appearance much like that of a hiker's contour map. Here the "highs" and "lows" of the atmospheric landscape are shown just as are the peaks and valleys of our terrestrial home. One can easily imagine the winds rushing down the "steep mountain slopes" to fill the valleys below (Figure 2.5).

Generally, where the "lines of equal pressure" wrap themselves around a low-pressure area, you will find them much closer together than those encircling an area of high pressure. (See Figure 2.5). This explains why lows are most often accompanied by higher winds: Their "hills" are simply steeper.

Winds Flow From High to Low

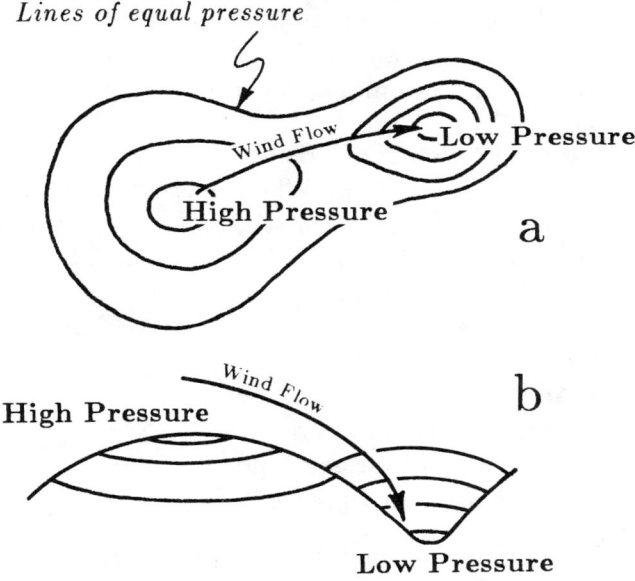

Figure 2.5: (a) Pressure differences occur most quickly where the lines of equal pressure are closest together. (b) Therefore, air moves fastest on the "steep" side where pressure differences are greatest over the shortest distance.

The Way of Moisture

> When warm, moist air is made to rise,
>   Clouds are made to fill the skies.
> When clouds are made to fill the skies,
>   Rain, she comes as no surprise.

There is much more to rising and falling air than just up and down motion. Where there is air on the rise, clouds are bound to fill the skies. And where there is descending air, clear skies abound. The way of moisture in the atmosphere explains why:

> When warm air rises, air expands,
>   And when it expands it gets cool.
> So the story is told,
>   when rising air gets cold,
> Clouds appear as a general rule.

> When air slowly makes its way down again,
>   It compresses, and that makes it warm.
> And where air is too warm,
>   a cloud cannot form,
> So clear skies appear as the norm.

Described here is what takes place in the rising and falling air of large-scale weather systems, as well as what occurs in and around an isolated cloud. As air rises into the atmosphere it expands, and expansion causes cooling. Cooling in turn causes moisture to condense. And, what do you know, when moisture condenses we find ourselves with a cloud (Figure 2.6).

The Way of Moisture 19

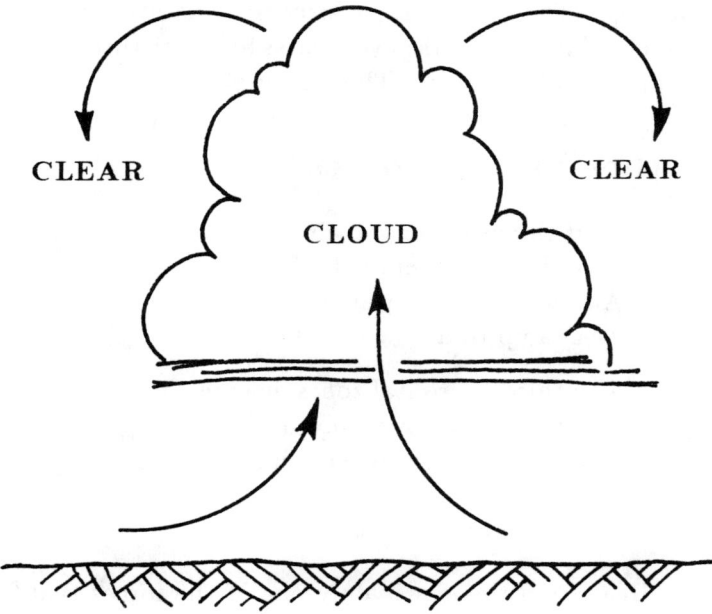

Figure 2.6: Rising air expands and cools, resulting in clouds. Sinking air compresses and warms, causing clouds to evaporate.

In the clear air all around a cloud, however, we have just the opposite taking place. The air here is slowly sinking, and as it moves downward, it compresses. This compression process causes warming which all in due course evaporates any moisture that happens to be present. Thus, clear sky.

Other Lifting Forces

> Four things cause the air to rise,
> And one of them is heat.
> Another is a mountain's sides—
> A ramp to any winds they meet.
>
> The third in frontal zones emerge
> When one mass climbs another's slope.
> The fourth occurs when winds converge
> And air squeezes upward
> like a wet bar of soap.

Heat from the sun is only one of the forces that works to push air aloft. In addition to heat, there are three other major lifting forces, all of which are capable of bringing about the same cloudy—and very often rainy—results:

1. A mass of air, when blown against the side of a much more solid mass as that of a mountain, has little choice but to ride upward on the mountain's slope (Figure 2.7b).

2. Likewise, a mass of cold air, being thick and heavy, acts very much like a mountain when

Other Lifting Forces

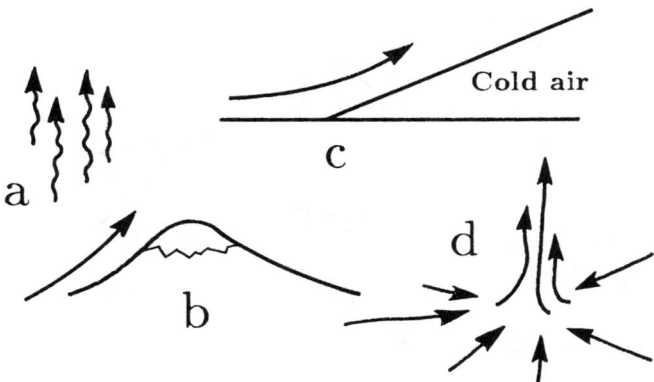

Figure 2.7: The four primary lifting forces: (a) heat, (b) mountains, (c) masses of cold air, and (d) converging winds.

encountered by a body of warmer, lighter air. When the two meet, a "front" is formed, and the warmer air rides right up and over the cooler air (Figure 2.7c).

3. And, finally, when winds blowing from different directions happen to converge on a common point, the air piling up as a result takes the only route open to it—upward motion (Figure 2.7d).

Remember, too, no matter what the lifting force happens to be—whether the sun, a mountain, a cold air mass, or converging winds—when air rises, it expands and cools. Due to this reduction in temperature, any moisture the air happens to contain is likely to condense into clouds or rain.

Figure 2.8: The windward side gets the clouds and rain, while the leeward is generally warm and dry.

The principle remains true on the other side of the mountain as well. Air that has already risen has likely shed a good amount of its moisture. Therefore, when it makes its descent again, not only is it drier than it was, but it also becomes steadily warmer due to compression. And there is little better combination for clear skies than dry, warm air. Thus, the windward side of a mountain gets the clouds and rain. To the leeward we find things generally warm and dry (Figure 2.8).

> Lakes and oceans add it, then
> Mountains wring it out again.

The cycle of the waters and the cycle of the winds are really one and the same. But it's all just a matter

Other Lifting Forces

of course for our weather. Heat and pressure keep the air constantly moving, always replenishing. What the winds remove, they replace. And what the weather takes, it gives again in return.

Chapter 3

Coriolis: The "Swirling" Force

> Coriolis, my darling,
> Coriolis, my dear,
> You say that our love spark
> Will ne'er disappear.
>
> And if e'er the pressures
> Of life should arise,
> The storm winds that strike us
> Will not steal our prize.

The "Swirling" Force

> Yet **high to low**
> And never returning,
> The winds have extinguished
> My heart's tender yearning.
>
> For try as I may
> And try as I might,
> I'm "always wrong"
> While you're **"always right."**
>
> Now don't be downcast,
> And don't feel remorse,
> For that's just your way
> Coriolis force.

There are two key phrases in this rhyme, both of which are emphasized in bold print. The first summarizes a principle covered in the previous chapter: Winds flow *away* from a high-pressure area *toward* an area of low pressure. Always high to low. (See Figure 2.5 on page 17).

Well, as you might guess, things of the atmosphere just don't go about their tasks so simply. Instead, what we find happening is that, as the winds are making their way from high to low, they are all the while being deflected *to the right* by a force known as "Coriolis force," and thus, key-phrase number two. (Bear in mind, this is how things work on the *northern* half of our globe—Figure 3.1).

This force—"Coriolis force"—has both global and local weather implications. First, the global:

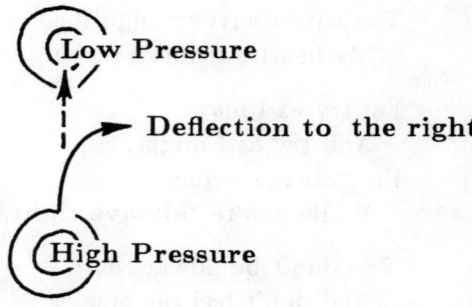

Figure 3.1: While the winds are making their way from high pressure to low pressure, the force of Coriolis is busily deflecting things to the right.

> Everyday, from the equator
>   straight to the pole,
> Our air would catch a regular flight,
> Except that as it passes midway,
> There's always a detour to the right.

This is Coriolis at work on a global scale. Because of the way the earth's surface is exposed to the sun, warm air rises at the equator and a pattern of circulation like that illustrated in Figure 3.2a seems the natural course to follow. Well, Coriolis force happens to think otherwise, and remember, Coriolis is "always right." The real-to-life circulation pattern we find traversing our world more closely approximates that shown in Figure 3.2b.

On a smaller scale, here's what Coriolis does to

The "Swirling" Force

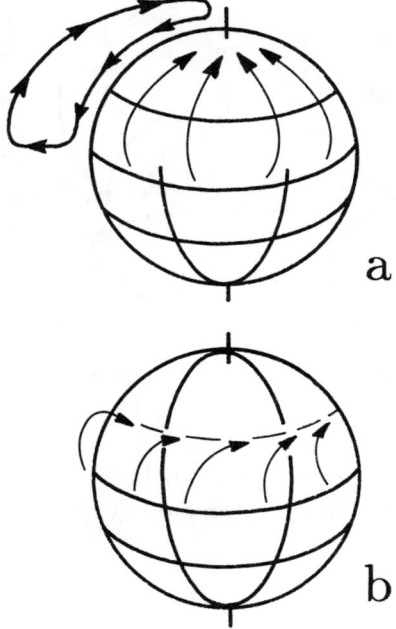

Figure 3.2: Instead of warm equatorial air heading straight for the poles (a), the force of Coriolis deflects the flow of air to the right just about midway in its travels (b).

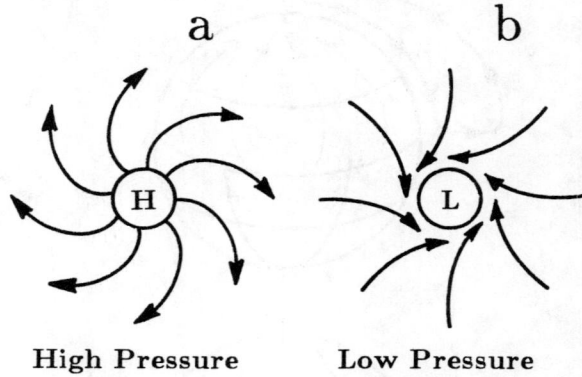

High Pressure   Low Pressure

Figure 3.3: Winds flow from high pressure to low, while the right-turning force of Coriolis causes them to swirl.

our winds: Air wants to flow from areas of high pressure to areas of low pressure. But because of Coriolis' constant push to the right, air flowing outward *from* the center of a high, rather than taking a straight path, is instead forced to take on a clockwise swirl (Figure 3.3a). On the other hand, air flowing *toward* the center of a low, because it is also being forced to the right, takes on a *counter*clockwise swirl (Figure 3.3b). Thus, Coriolis' nickname: The Swirling Force.

The "Swirling" Force

>Tighten a bolt,
>>Increase the pressure.
>
>Loosen a screw,
>>Pressure is lesser.
>
>Tighten it swiftly
>>For fair-weather wind,
>
>Loosen it quickly
>>Fierce storm to begin.

Knowing and understanding the direction of air flow around areas of high and low pressure is an important tool for weather analysis and prediction. (You'll see why in the chapters ahead). And the above rhyme is a handy one for remembering which way the winds swirl with which pressures. When you tighten a bolt, what do you do? Two things: You cause high pressure, and you rotate the bolt clockwise (Figure 3.3a). See how the two go together? The opposite happens when you loosen a bolt. The pressure becomes low, and the turn is counterclockwise (Figure 3.3b). Simple as that!

Chapter 4

Global Winds and General Air Circulation

The Westerlies

>Facing west, you can almost bet it,
>Whatever you see, you'll probably get it.

While that may sound a bit pessimistic, one has to remember, weather doesn't consist only of drizzle and cold. If a big gray blanket of clouds should appear in the west, yes it's likely headed your way. But if cheery patches of blue sky should appear, take heart, they're just as likely to let in a little sun.

But there's good reason for such probability. You'll recall that the sun's constant beating on the equator causes air to rise from that region of the

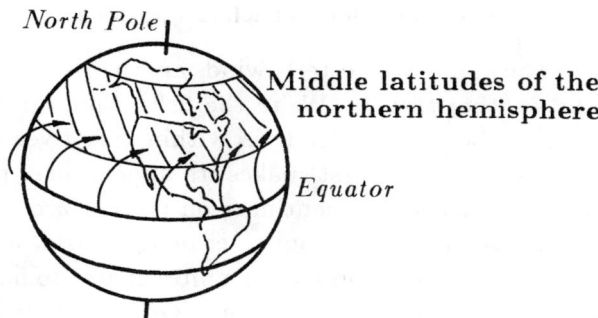

Figure 4.1: The "stream bed" of the upper-air westerly winds.

earth's surface.[1] Once it has risen into the atmosphere, the air pushes its way toward the coolness of the poles. In the case of the northern hemisphere, the air would keep right on traveling northward except for the strong right-turning force imposed by Coriolis[2] which turns the flow of air away from its original south-to-north path until it settles into its west-to-east track high above our middle latitudes. These latitudes, which form the "stream bed" for the river of the upper-air westerlies, span from the Gulf Coast to the southern portion of Alaska (Figure 4.1).

---

[1] For more information on the principle of rising air, see the first section of Chapter 2.

[2] For a discussion of Coriolis and its effects, see Chapter 3.

> To get a glimpse of tomorrow's theme,
> Just take a look at what's upstream.

The upper-air westerly winds are the river in which our weather travels. We're the fish. And our homeland is the river bed. It is within this river of air that our weather first makes its appearance, passes over our heads, and continues on its way around the globe. Of course, with our river being as wide as it is, none of us is so fortunate (or unfortunate) to have *all* the weather pass right over us. On the other hand, the principle is clear: Our weather in the middle latitudes of the northern hemisphere moves in a *generally* west-to-east direction.

> The westerly currents
> Will blow just about
> Five-hundred miles in a day,
> So don't doubt;
> That when you look west
> Or east you can say:
> "It's coming from there,"
> Or, "It went that-a-way."

The flow of our weather river is from west to east. And the speed of its current is about 500 miles a day. From this it's easy enough to see that the weather presently lying about 500 miles to the west of you will likely be your weather tomorrow. So, keep an eye on things to the west!

> Why the rains of spring and fall?
> Where are the westerlies?—
> They'll tell it all.

The Westerlies

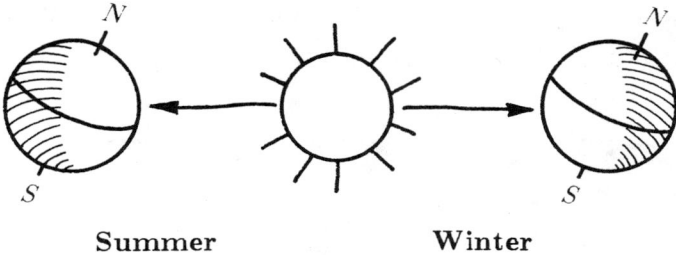

Summer   Winter

Figure 4.2: The tilted axis of the earth, together with the rotation of the earth around the sun, gives us our seasons of warmth and cold.

As the earth makes its annual trek around the sun, the tilt of our globe on its axis causes the rays of the sun to strike its surface at varying angles (Figure 4.2). Of course, the more direct the sunlight, the warmer the earth on that particular band of the earth's surface. During our winter, the sun is shining directly on the southern latitudes, and not so brightly in the north. In our summer, just the opposite happens.

Because of the tilted axis of the earth in relation to the sun, the earth's "belt of greatest warmth" is constantly shifting up and down on its great belly. And shifting right along with this "belt of warmth" are all the other forces that make our weather such a fascinating phenomenon to study and observe—

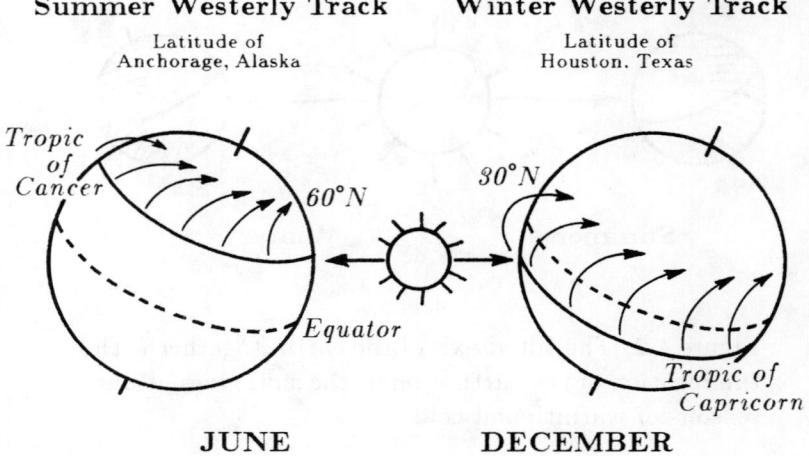

Figure 4.3: The seasonal migration of the westerly winds.

among these, the right-turning force of Coriolis [3] and the resulting westerly stream of upper-air winds (Figure 4.3).

> Hup, two, three, four:
>   The westerlies are at my door.
> Seven, eight, nine, ten:
>   The winds are heading south again.

The high-altitude westerly winds seem always to form in a band that circles the earth an average of 30

---

[3] For a discussion of Coriolis and its effects, see Chapter 3.

degrees to the north of where the sun happens at the time to be shining most directly on the earth. Because of this, as the sun moves north and south with the changing seasons, so does the westerly stream (Figure 4.3). The upper-air westerlies are in essence the river in which most of our weather systems flow. Thus, as the river of westerly winds makes its seasonal passage over our heads from south to north and back again, so does the stormy weather which typically flows in its stream.

"Hup, two, three, four"—the fourth month of the year is April, the approximate time of year we can expect the mid-stream of the westerlies to be passing overhead, and thus the month of showers. "Seven, eight, nine, ten"—October is the approximate time of year when showers again are frequent as the westerlies pass over our heads on their way south (Figure 4.3). It is the position, or "track", of the westerly stream, then, that determines just where the storm belt lies at any particular time of the year.

Seasons

>Winter, spring,
>>Summer, fall,
>There is a way
>>To remember them all.
>
>Twelve, three,
>>Six, and nine;
>Add a "slash" after each,
>>Then to each assign:

Twenty-two, twenty-one,
   Twenty-one and twenty-two,
Now watch what this little trick
   Does for you.

Winter, spring,
   Summer, fall,
You now have the dates
   To start them all.

From the above rhyme you can see that the exact day for each season's start varies between the 21st and the 22nd of the month in which the seasonal change occurs. The memory aid is offered here to help you remember which day corresponds to which season, as well as to keep track of the shortest, longest, and two "equal" days of the year. Take a twelve, a three, a six, and a nine, then add to each the appropriate "slash twenty-one" or "slash twenty-two." Here's what you get:

- **12/22** for the start of **winter** (shortest day of the year—when the sun is shining directly on the Tropic of Capricorn in the southern hemisphere).

- **3/21** for the start of **spring** (spring equinox—when the daylight hours equal the nighttime hours).

- **6/21** for the start of **summer** (longest day of the year—when the sun is shining directly on the Tropic of Cancer in the northern hemisphere).

- **9/22** for the start of **fall** (autumn equinox—when the daylight hours again equal those of the night).

Knowing these dates can also help you in keeping informed of where the westerly stream happens to be at the time, and of how its storm-bearing winds are progressing in their north-and-south migration. At the start of winter, you'll find them in their most southerly position (in the neighborhood of Houston, Texas—about 30 degrees latitude north). By the start of summer they will have made their way to their most northerly position (in the vicinity of Anchorage, Alaska—about 60 degrees latitude north).

Air Masses

> As seasons pass,
>   And it's time to reminisce,
> The oceans retain,
>   While the lands dismiss.

Our oceans, you might say, are the "shock absorbers" for our earth's constantly changing temperatures. Because they're slow to change their temperature, the oceans tend to have a moderating effect on the much faster and more extreme temperature fluctuations being experienced by our land masses. These characteristics of water and land are essentially the how and why of the air masses that cover our earth—those great bodies of air characterized particularly by their similar properties of temperature and moisture.

> Dampness and dryness,
>   and hotness and cold,
> An air mass has generally
>   two of these.
> Originating south makes it hot,
>   and north cold;
> Dryness from land,
>   and dampness from seas.

And what do these air masses have to do with our weather? Well, picture if you will a huge, jello-like mass of bitterly cold, dry air moving very slowly through your neighborhood. Because of its great expanse, it takes several days to make its passage, all the while enveloping everything in its path with its frigid, numbing cold. A different mass of air, on the other hand, may be traversing another part of the country, bringing with it muggy, warm air to add humid misery to an already hot, but normally dry climate. Both are referred to as air masses, and the weather they bring is known as air-mass weather. Their primary difference, however, is in their place of origin (Figure 4.4).

There is a reason why all these air masses get to be the way they are in terms of temperature and moisture content. Because they develop in areas that lend themselves to keeping air around a long time, they have little trouble building a sizable mass of hot, dry air, for instance, or cold, dry air, or cold, damp air, and so on. The masses themselves, then, tell us quite a bit about the originating areas and their climates.

Air Masses

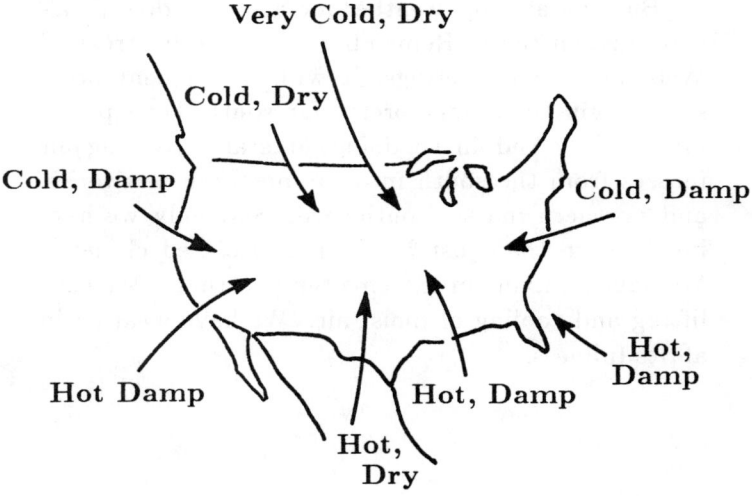

Figure 4.4: Air masses are characterized by similar properties of temperature and moisture.

> Definitions for "climate"
> All put together,
> Usually end up
> Meaning "average weather."

But the average weather for a region doesn't always just sit there. Remember the westerly stream[4]? Well, in its meanderings, it will on frequent occasions begin to wander pretty far south, then pretty far north. And in so doing, it starts to drag air masses from the south into the northern territories, and northern masses southward. Suddenly we have much more than just a bunch of isolated climates. We have a mixing of extreme temperatures. We have lifting and cooling of moist air. We have weather in all its fullness.

---

[4] For a review of how the westerly stream of upper-air winds is formed, refer to the first section of this chapter.

Chapter 5

Weather System Winds

Surface Friction

> 'Round and 'round the weather winds blow,
>   And where they stop you'll never know,
> For if they're blowing north today,
>   Tomorrow they may blow another way.

So far we've talked a good bit about the upper-air westerlies[1] —those high-altitude winds that flow from west to east over our land and carry along in their stream our various forms of weather. But now it's time to get down to earth.

We live right in the stream bed of the westerly winds. And knowing that, a likely question follows: If the westerly winds are constantly blowing over our

---

[1] For a thorough discussion of the upper-air westerly winds, their formation and their role in our weather, see Chapter 4.

heads from west to east, why is it that the wind we see blowing rain clouds over our gardens is coming from the south? And why does the wind then reverse itself and bring clear skies out of the north? The answer lies in the circular nature of our weather-system winds. And friction is the cause of it all.

> What friction does to Coriolis
> Is like having a cart
>   but nothing to pull us.
> 'Cause slowing the wind
>   weakens Coriolis' force,
> Just as sticky ol' mud
>   would weaken a horse.

Wind wants to flow from high pressure to low pressure. Then in the midst of the wind's travels, Coriolis[2] goes to work turning things to the right. And that's what sets up the high-altitude westerlies.

But high-altitude winds are different from those down here closer to the earth's surface. The winds way up there don't have anything to rub against— nothing to slow them down. Down here near the surface the winds have mountains and trees and all kinds of other things to stand in their path. The result is friction (Figure 5.1).

If friction did nothing more than slow the winds, you would think that here on the ground we should just have weaker westerlies blowing. Well, friction happens to do more than just slow things down. In

---

[2] For more information on Coriolis force and the formation of the high-altitude westerly winds, see Chapter 3 and the first section of Chapter 4.

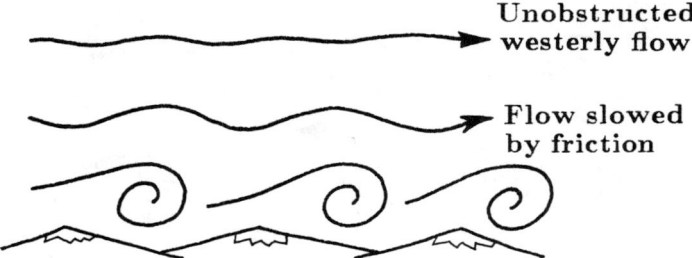

Figure 5.1: Air flow near the earth's surface is slowed by friction.

slowing things down, friction produces another very significant effect: It also weakens the right-turning force of Coriolis.

Let's take an isolated low-pressure area to illustrate what's taking place. Because of the way air responds to pressure differences, winds try to blow toward the center of the low (Figure 5.2a). Coriolis meanwhile turns things to the right (Figure 5.2b). But—and here's the clincher—the force of Coriolis' right-turning abilities is dependent upon how hard the wind is blowing. Way up in the atmosphere where friction is negligible, Coriolis has a lot of clout and the right turn made by the wind is therefore pretty sharp—thus, the westerlies. Get down here closer to the earth's surface, though, and things begin to change. Friction starts to slow the wind, and Coriolis can't turn things quite so far to the right. As a result,

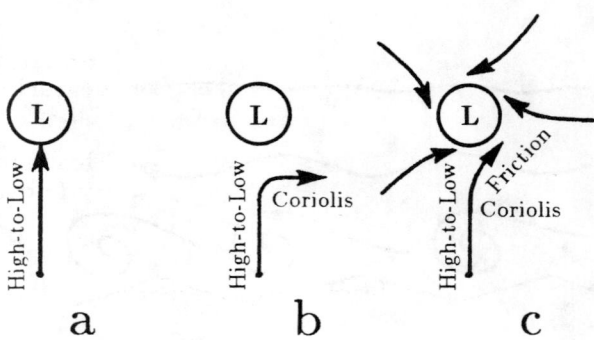

Figure 5.2: Surface friction causes winds to *spiral* toward an area of low pressure.

the winds neither flow directly toward the low, nor do they turn sharply to the right. Instead, the winds flow in a gentle *spiral* toward the low (Figure 5.2c).

Spiraling Winds and the Westerly Stream

> On the average, a weather system will go
> Five hundred miles a day or so.

The spiraling winds of our weather systems are much like spinning leaves being carried along in a rapid brook. Though our weather systems first spin this way then that, they continuously flow from west to east in the giant stream of the westerlies. The winds *within* the systems flow in circular patterns due to friction's effect on Coriolis' force (see Figure 5.2),

Spiraling Winds 45

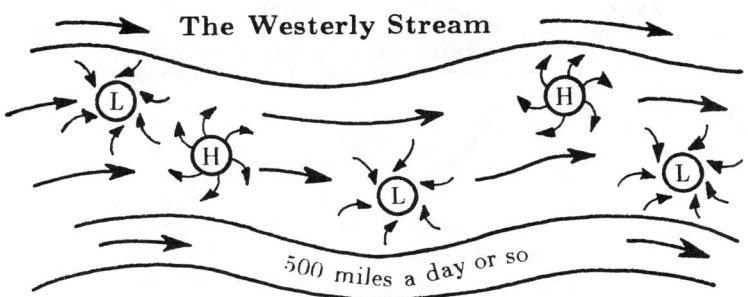

Figure 5.3: Systems of high and low pressure, together with their spiraling winds, are carried along in the current of the westerly stream.

and because of this, we experience the shifting of winds as the various sectors of a storm pass over us. As the winds are busily spiraling this way or that, each individual spiraling weather system is all the while being carried as an intact weather system down the westerly stream (Figure 5.3).

> A circle has
> Two halves, alas,
> So slow to come
> Means slow to pass.

Systems of high and low pressure are somewhat like giant circles moving across the land, carrying within their boundaries good or bad weather as the case may be. Being generally symmetrical in shape

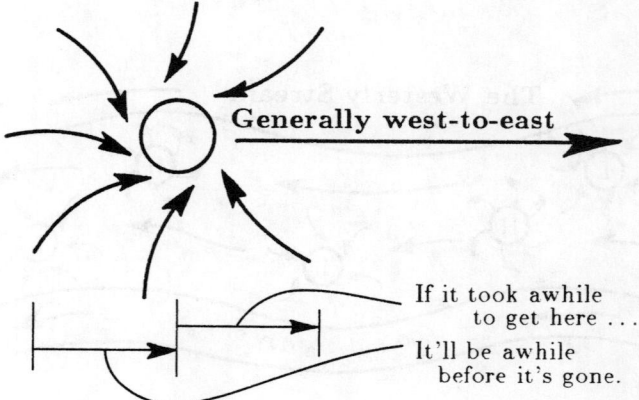

Figure 5.4: Weather systems of high and low pressure are a bit like giant circles moving across the land. Their symmetrical nature causes them to take about as much time in leaving as they do in getting here.

means this: If a storm center is a long time in reaching you, it will also be a good while before you see an end to it (Figure 5.4).

> The smaller the high,
> The faster gone by.

The principle here is the same as the one described in the previous rhyme. And it just goes to show that with a little understanding of the rules that govern our atmosphere, our weather begins to make sense. Principles start to relate. Rules begin to overlap. Cloud movements and wind direction suddenly have something important to tell us. And for the most part, all we have to do to receive their message is simply to keep looking up!

Chapter 6

Mentally Positioning Yourself Within a Weather System

The Wave Cyclone

    Hands out front, thumb-to-thumb.
        Pinkies down and then some.
    Thumbs are center of the low,
        Fingers point to frontal flow.

    Fingers on the left are cold,
        And in your sweaty palms you hold,
    Warm moist air that pushes east
        To drop its rain on man and beast.

    Provided here is a visual aid to help you depict

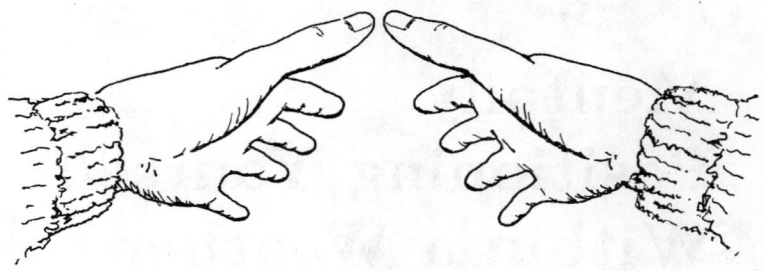

Figure 6.1: Using your hands as a visual aid, you can depict the typical wave cyclone as it moves as an integral whole across the land. For specific instructions, refer to the first rhyme of this chapter on page 47.

in your mind's eye the typical wave cyclone as it moves as an integral whole across the land. (Refer also to Figure 6.1). A cyclone, by definition, is any mass of air that happens to be swirling in a counterclockwise direction. This is characteristic of a low-pressure system and is the opposite of an anticyclone which, you guessed it, rotates clockwise and characterizes an area of high pressure.[1] You can remember all this by simply noticing that cyclone has

---

[1] To help you keep it straight which direction the winds swirl when systems of high or low pressure pass by, refer to the closing rhyme of Chapter 3 on page 29.

The Wave Cyclone

two "c"s at its start. Think of these as standing for counterclockwise. Then just remember that if you add the prefix "anti" you get the opposite. And there you have it!

Now that you know what a cyclone is, I'll bet your wondering, "What on earth is a *wave* cyclone?" By following the instructions of the rhyme on page 47, our outstretched hands form an inverted "V" with the line of their finger tips (Figure 6.1). It is this inverted "V" structure along with the temperature and moisture characteristics of the air both inside the "V" and all around its outside that will make our weather understanding suddenly begin to soar. But first, let's see just how this wave gets its start in the first place.

We've talked quite a bit now about how the westerlies are formed by air being deflected to the right as it travels north from the equator. Well, not *all* of that air gets deflected. Some of it slips through Coriolis' fingers and makes its way on up to the polar regions of our globe (Figure 6.2a). As the air approaches the pole, it sinks to the surface and starts its slow trek southward, all the while becoming very cold and heavy. When it finally reaches the latitude at which the westerly stream is flowing, it begins to pile up. It is this "pile up" region that we commonly refer to as the polar front, with cold air to the north, and warm(er) air to the south (Figure 6.2b). This is the line along which our wave cyclones form.

The whole formation process of a typical polar-front wave is stepped out in Figure 6.3. Things get started with just a ripple in the polar front—a

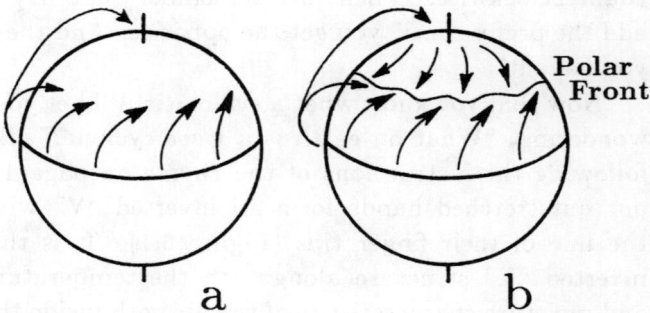

Figure 6.2: (a) As air travels north from the equator, some of it eludes the right-turning force of Coriolis and continues on to the pole. (b) Once at the pole, the air sinks and slowly makes its way south until it piles up in the region of the westerlies where the polar front is formed.

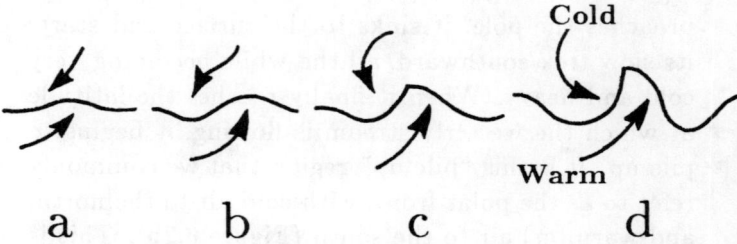

Figure 6.3: Illustrated here is the step-by-step formation process of the typical polar-front wave.

The Wave Cyclone

little air pushing north, and a little air pushing south (Figure 6.3a-b). Soon the ripple turns into a swell (Figure 6.3c). Then it becomes a full-fledged wave (Figure 6.3d).

As you can see in Figure 6.3d, cold polar air to the left of the inverted "V" is pushing south and eastward. Thus we find ourselves with a cold front. Inside the inverted "V" and over to the right we find warmer air pushing eastward and north, causing the formation of a warm front. Suddenly we have before us a typical wave cyclone—also known as a "polar-front" storm.

Keeping your hands held as shown in Figure 6.1, move them to the right in a broad sweeping motion. Now picture yourself huddled somewhere on the earth's surface in the path of this moving storm. With just a bit more understanding of the characteristics of warm and cold fronts and what to expect in the various quadrants of a storm, you will find yourself well equipped to read the sky and the winds. You will be able, with a fair amount of accuracy, to position yourself in relation to a storm and to know what kind of weather to expect.

> Wind at your nose,
> Arms stretched like a crow's;
> Storm center to your right,
> And over your shoulder a mite.

Low-pressure weather systems of the northern hemisphere always rotate in a counterclockwise direction, spiraling about the center of a storm. Thus, as you stand with the wind in your face, the storm center

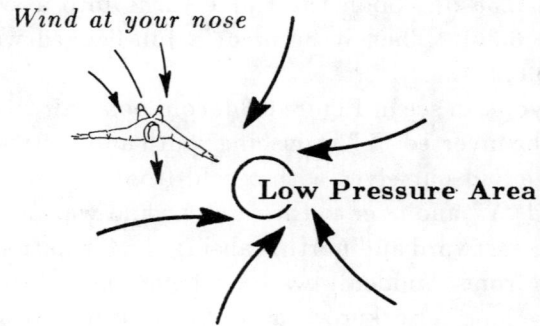

Figure 6.4: To locate the center of a low-pressure area, face into the wind and stretch out your arms. The storm's center will be to your right and slightly behind you.

will be found to your right. However, because winds tend to *spiral* toward the center of a low, the storm center is actually to your right and slightly *behind* you as you face into the wind (Figure 6.4). Clouds can be used to confirm your findings as they are generally thicker and lower toward the low-pressure area.

Warm Fronts

>Warm front creeps smoothly
> Up over the cold,
> By the fringe of its blanket,
> Long foretold.

As a warm mass of air encroaches on an air mass colder than itself, rather than doing battle eye-to-eye,

Warm Fronts

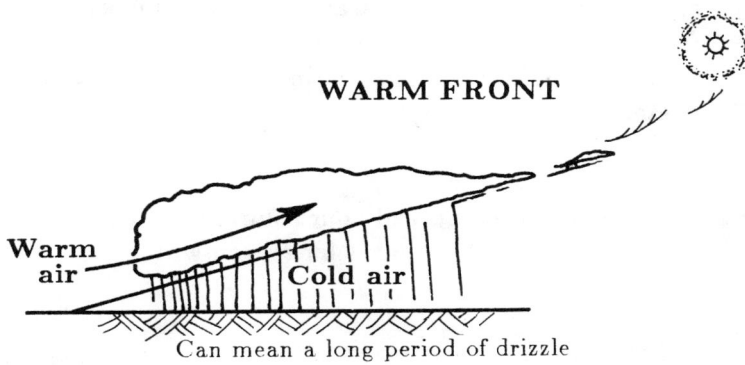

Figure 6.5: Shown here is the cross section of a warm front. Note the very gradual lifting of moisture and the wide band of precipitation that results.

the warmer air begins a smooth upward journey as it overrides the cooler air as if ascending a long, gently inclined ramp (Figure 6.5).

As the warm, moist air travels skyward, of course, it expands, cools, and condenses into the expansive blanket for which the warm front is well known. The extreme leading fringe of this blanket may be nothing more than an ever-growing halo around the sun.[2] Horse tails of cirrus[3] then give a more obvious clue,

---

[2] For right now, note where it is that a halo occurs within the overall weather system. For more on halos, check the index.

[3] Note where it is within a storm system that these wispy clouds generally occur. For more details, check the index.

and when these in turn lead to a steady thickening and lowering of the cloud deck, you can be pretty sure a low is well on its way. Studying Figure 6.5 from right to left will give you a blow-by-blow account of what to look for during the approach and passage of a warm front.

> Warm front spread your colorless quilt.
> Warm front spread it far as you wilt.

The gradual slope of the warm front, which may extend as many as several hundred miles, can spell drizzle for days on end (Figure 6.6).

> The warm front sends criers far ahead
> To make itself early known.
> The cold front will lean over backwards instead
> To hide its presence until it has blown.

As a warm mass of air overtakes a mass of cold air, the warm air starts a long, steady climb over the cold. The earliest warning the warm front gives to those in its path is therefore the wispy cirrus horse tails that are high up at the top of the slope, telling well in advance that, though the base of the front is yet a good way off, it's on its way nonetheless. (See Figure 6.5). A cold front, however, does just the opposite.

Warm Fronts

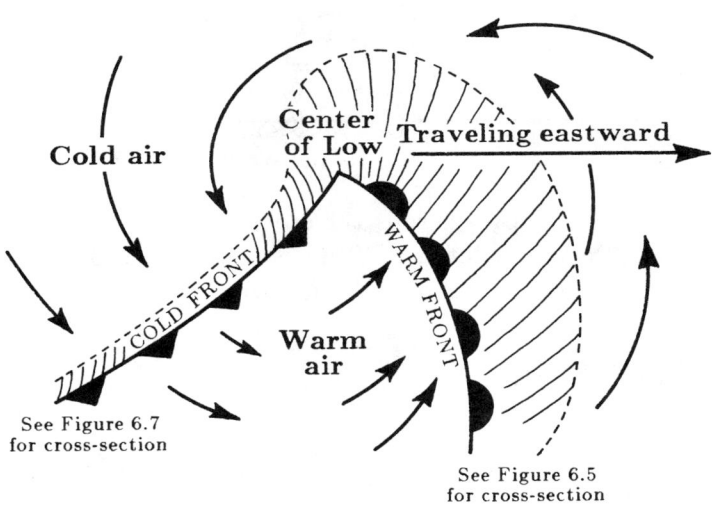

Figure 6.6: This top view of a typical wave cyclone shows the cold and warm air masses, and the cold and warm frontal systems. The shaded areas indicate where rain is most likely to occur.

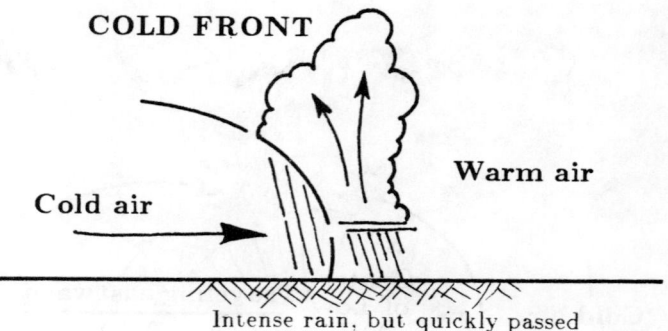

Figure 6.7: Shown here is the cross section of a cold front. Note the intense lifting of moisture and the relatively narrow band of precipitation.

Cold Fronts

> A cold front rolls
> Like a bowling ball,
> And sometimes more like
> A speeding brick wall.

Cold fronts are typically characterized by their relative viciousness when compared to their warmer counterparts. The reason is really quite simple. A cold air mass pushing against a warmer mass of air forms a sharp vertical wall up which the warm air ahead is forced to climb (Figure 6.7). In fact, the cold front tends to lean even a little bit backwards. There are no high-altitude early warning signals here. When you can see the cold front, the cold front is here.

> Cold, thou art;
> Cold front, cold heart.

Such an abrupt climb can cause a good bit of violence. Think of it: Cold air advances like a dull but very powerful wedge. And the faster it moves, the more violently it forces warmer, more humid air high into the atmosphere. It's little wonder that heavy showers and gusty winds hail its presence (in more ways than one). The abruptness and speed of the cold front are also what make its band of precipitation narrow and its winds short lived. Study Figure 6.6 and compare the frontal systems indicated there with those diagrammed in Figure 6.5 and Figure 6.7. Fixing this picture in your mind of what occurs where within a typical storm structure will aid you greatly in your efforts to analyze and predict the weather.

Identifying Frontal Passage

> Temperature changed.
> Clouds rearranged.
> Wind shifted.
> Humidity drifted.

These are the four points by which you can identify the passage of a frontal system—warm or cold. Used together they make a pretty reliable team for determining whether the atmospheric stuff that has just occurred is somehow related to a frontal system or whether there is some other perhaps more localized cause for what has taken place.

OCCLUDED FRONT

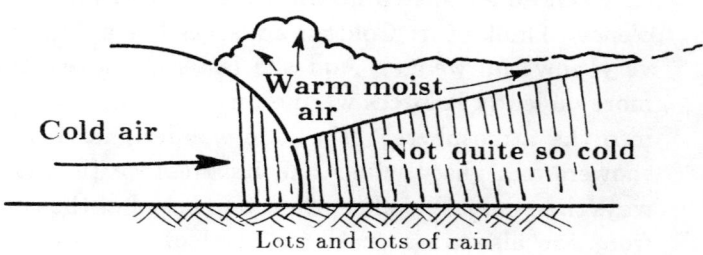

Figure 6.8: Pictured here is the cross section of an occluded front. Note how moisture has been trapped aloft between the cold and warm fronts. This generally results in a prolonged period of intensified rain.

Occluded Fronts

> A frontal occlusion
> Means rain in profusion.

Cold air tends to move faster than warm air. With this fact in mind, apply the "hands out front" rhyme that started this chapter, and envision how things might proceed. If the cold front keeps up its faster pace, it will soon overtake the warm front. When this happens an occlusion occurs. Essentially, this means a pocket of warm air has been trapped and thrust aloft where the cold and the warm fronts have come together (Figure 6.8).

Two things result. First, due to the more intense and widespread lifting of moist air found in an occlusion, the drizzle generally intensifies into a good steady rain. And second, the occluded part of the storm (once it *becomes* occluded), likes to slow down, making the now intensified rain hang around a good bit longer than it otherwise would have. Again, using your "hands out front" rhyme, you can see that the most likely place for occlusion to first occur is right at the top of the inverted "V" where your two hands come together. Therefore, the real widespread, steady rains of a storm are found most often just to the north of the storm's center (Figure 6.6).

Storm Quadrant Characteristics

> A storm passing south
>   Brings clouds, and cool,
> And drizzle as steady
>   As a three-legged stool.
>
> A storm passing north,
>   Though starting with rain,
> Will soon warm, then shower,
>   Then dry up again.

We have just learned what sort of weather accompanies warm, cold, and occluded frontal systems. Now we're ready to review just what it is that happens between and around these fronts. Hold your hands up again to get the picture. (See Figure 6.1). Sweep them to the right now as if the storm was traveling east. Now imagine yourself to the north of this weather system as it goes merrily sweeping by. What sort of weather do you see pass overhead? Well, take a look at Figure 6.6 and mentally position yourself within the diagram. What you're likely to experience is the steady drizzle dropped by the northern portion of the storm. Take note how this analysis, together with a quick check of the wind's direction in order to test for the location of the storm's center[4], gives you a pretty firm indication that the storm is passing by to the south of where you are.

But what if you should find yourself to the south of the storm's center as it passes by? Try it. (Refer again to Figure 6.1 and Figure 6.6). First the horsetail fingers of the warm front's leading edge flow over your head. The clouds slowly lower, and a steady rain sets in. As the warm front passes, skies clear and the air turns warm(er) and muggy—perhaps unseasonably so. Before long, however, winds start to gust, clouds begin to tower in the sky, and a brief period of showers is followed by cold, dry air under blue skies with the passing of the cold front. As you can see, the southern portion of a storm can be a bit more violent (though perhaps a bit more interesting, of course) because of the passing fronts. This is especially true if temperature differences are extreme.

Remember, a storm has structure, integrity. This whole structure, as described above, moves as a unit across the country. Not only does the air move within the system, but the whole system moves.

And remember, too: What we have just learned

---

[4]For a description of this simple test, refer to the rhyme on page 51 and take a look at Figure 6.4.

Storm Quadrant Characteristics

concerns the "typical" wave cyclone. It serves as an excellent guide to understanding what's going on in the atmosphere around us. But be on the lookout for all sorts of variations. Nature does have its way of doing things just a little differently each time—and sometimes a *lot* differently!

> Beware! Beware!
>   The winter storm,
> Some parts cold
>   And some parts warm.
>
> To north of the center
>   You'll find much snow,
> A comfort to only
>   The Eskimo.
>
> And to the east
>   Will be floods and ice,
> Rain on frozen ground
>   Is not so nice.
>
> To the south you'll experience
>   Unseasonable thaw,
> But it's only a part
>   Of the weather seesaw.
>
> For yet to the west
>   Looms a wind and a chill,
> That'll sting to the bone,
>   And may even kill.

Here we have the winter version of the wave cyclone. If you take all that you've learned so far about

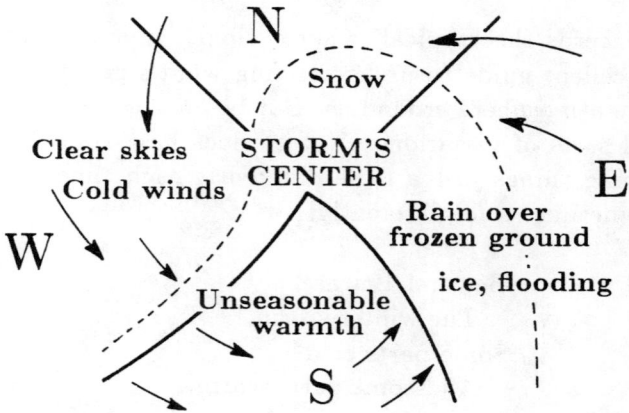

Figure 6.9: Weather characteristics found in the four quadrants of the winter storm.

this type of storm and just add a good frightful chill, this is what comes out. (See Figure 6.9). Notice that the steady rain to the north has turned to snow. It still rains to the east, but with harsher consequences. The south quadrant (inside the inverted "V", remember) is still the relatively warm sector. But the west, because of the bitter cold drawn from up north, can be a real chiller—and with a little wind chill thrown in, even a killer. So watch out!

The Eddy Cyclone

> Little brother Eddy
> Flows a lot more steady.

Though we've been dealing chiefly with what is known as a typical wave cyclone, the characteristics

The Eddy Cyclone

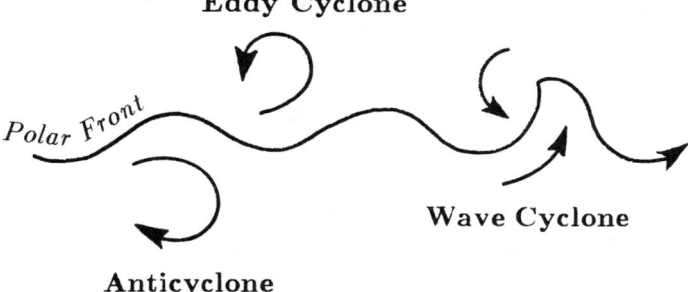

Figure 6.10: The eddy cyclone behaves much like the wave cyclone as it crosses the countryside, but it lacks the sharply defined frontal systems.

and principles discussed in this chapter have given us the orientation needed to understand the passage of simple "eddy" cyclones as well. These eddy cyclones behave similarly to the wave cyclone as they traverse the countryside, but are so named because they aren't accompanied by the sharply defined frontal systems that the polar front provides the wave cyclone (Figure 6.10).

In these not-so-prone-to-one-extreme-or-the-other eddy storms, southern air is still transported northward, and northern air still travels southward. But here's the real difference: Because all this passing back and forth of air takes place within a single air mass, contrasts in temperature and moisture are small and the eddy storm just can't muster the violence that so often accompanies its bigger brother, the wave cyclone.

Chapter 7

Wind

Wind Basics

> Every wind has a story to tell,
> Adventure, high drama—or a wicked spell.

Not only is the wind a pretty decent story teller, it's one grand *fore*teller as well—of the weather, that is. By testing the wind to see just what it's up to, we can discover with fair certainty whether the stuff sitting out there to the west of us is an area of high pressure that wants to offer us a little sunshine, or a low-pressure area looking for a place to drop a little rain. We can predict the change of weather from good to bad. We can locate a storm's center. We can monitor its progress. And, of course, we can tell when a storm has finally decided to be moving along. All by the wind. But first, here are some basics for sharpening your senses for what the wind would have you hear.

> Trust a strong wind over a weak wind,
> A steady wind over a gust,
> An early or late wind more than midday,
> And watch the low clouds, not the dust.

What you see or feel should not be taken without question to be a sign from the heavens of what is to befall. In fact, some winds can be downright misleading. Weak winds, for instance, can be awfully fickle, meaning they just can't make up their minds where it is they're coming from or where they're going. A strong wind is much more trustworthy.

Gusty winds are a bit like weak ones. They tend to first blow from here then from there, never giving you any firm commitment as to what their real direction in life might be. So if a gust is all you have, cross your fingers and make a guess. Otherwise, put your feelers out for a steady wind—a much more reliable source on which to base a forecast.

There's also the time of day to be considered. When the sun has finally risen to its full height in the middle of the day, its rays have a way of producing localized hot spots. These spots of warmth then start to generate breezes which, if not recognized as such, can cause errors in your weather analysis. Things are generally a bit more settled earlier or later in the day, making these much better times to test and make judgements.

Then there's the matter of any local obstructions that might be producing swirls and eddies here on the ground. These "little" winds could very easily be blowing exactly opposite the winds that are pushing the clouds along just a few thousand feet above your head. So keep an eye up there, too.

Speed Indicators

>At ten leaves stir,
>>Weather vane turns;
>
>Small flags sigh,
>>Few concerns.
>
>At twenty limbs wave,
>>Newspapers blow;
>
>Leaves flutter about,
>>Small trees can't say no.
>
>At thirty trees sway;
>>The wind you admit,
>
>Is not leaning on you—
>>You're leaning on it!

Now here are some real-live observable indicators to aid you in making estimates of just how hard or not-so-hard the wind is blowing. You may like to add items to this list of indicators that are commonly present when you make your own weather observations. Observe, for instance, what the approximate wind speed is that makes a particular tree branch swish against the side of your house, or cause your clothing to be blown a certain way. In no time at all, your senses will be telling you with surprising accuracy what sort of wind is blowing and whether it is dying down or picking up.

>Past a hand width held high
> to a low-level sky,
>Clouds travel 'bout *thirty*
> if they pass in a minute.
>But a high-level sky
> passing two times as high,
>Takes a minute
> to do *sixty* within it.

The "thirty" and "sixty" here indicate miles per hour. Next time you're gazing up at some of your favorite clouds, give it a try. Looking straight up,[1] hold your hand at arm's length over your head to serve as your "measuring stick." Keep your fingers together and measure across the back of your hand. Now watch closely as you count off the seconds. The final outcome of your reading, of course, will depend on the height of the clouds being used for your wind-speed test, and the actual amount of time it takes them to travel the width of your hand. With a little practice you'll have this handy guide forever at your finger tips.

---

[1] Looking straight up gives you the truest reading by eliminating the distorting effects of perspective.

The Upper-Air Winds

> When upper-air westerlies
> Look like a trough,
> Look out 'cause the next storm
> May not be far off.

The high-altitude westerly winds[2] that blow overhead aren't something we can normally "see." But take a closer look next time you see horse-tail clouds weaving their strands in the sky. These cirrus clouds are formed so high in the atmosphere that they are actually being swept along by the westerlies. Needless to say, they serve as very effective wind-direction indicators when it comes to trying to figure out what the westerlies are up to. They're like wind vanes placed conveniently way, way up.

The westerly flow of air, being the meandering river that it is, doesn't always flow directly toward the east. In fact, sometimes it goes so far as to head more north or south. Things then begin to look more like the diagram in Figure 7.1. The stream still travels *generally* west to east, but in a very wavy fashion. And this is where high-altitude ridges and troughs begin to show themselves.

High-altitude air flow along these ridges and troughs is moving very, very fast.[3] Notice in Figure 7.1 that as the winds make their southward plunge

---

[2] For more information on the westerly winds, see the first section of Chapter 4.

[3] The swift winds that make up the air flow along these ridges and troughs is also known as the jet stream.

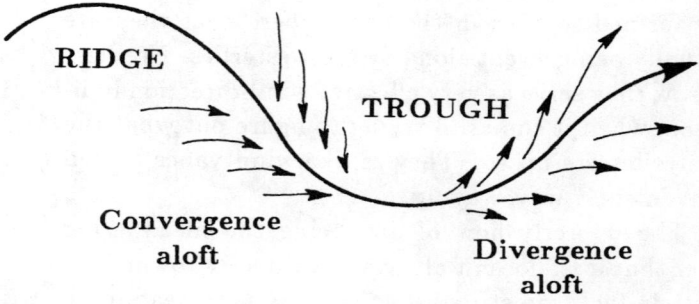

Figure 7.1: Looking down on the earth's surface as if from outer space, the meandering westerly stream takes on a wave-like pattern of ridges and troughs.

The Upper-Air Winds

down the side of a ridge, they also start to converge. The consequence of this convergence is a general pushing downward of the air below. This, in turn, induces (or at least encourages) an area of high pressure, which is accompanied, of course, by generally fair weather. When the upper-air winds again make their upward swing, things take a turn in more ways than one. Where convergence occurred on the downswing, the upswing brings about divergence, or outflow. It is this outflow aloft which, when strong enough, draws air upward from the surface. And before you know it, we have the start of a low, and a storm is born!

You might note, too, that the flow of air along the upper-air ridges and troughs explains why it is that areas of high pressure travel a generally northwest-to-southeast path, while low-pressure areas will generally move up the map on a southwest-to-northeast track.

> If from the south the horse tails fly,
> Clouds may surely fill the sky.
> If from the north the tails should scurry,
> 'Bout clouds and rain you needn't worry.

Because of the high altitude in which they form, horse-tailed cirrus clouds are pushed along by the upper-air westerly winds. This is what gives these clouds their special ability to indicate upper-air ridges and troughs. Here's how they do it: If the wispy wind vanes formed by cirrus clouds indicate air flow from the south, you are likely sitting on the eastern edge of an upper-air trough. Study Figure 7.1 closely. The dipping then rising motion performed by these high-altitude winds enhances the likelihood of a counter-clockwise twist and lifting of the winds below. This would be the birth of a low-pressure system which, in turn, would bring on the clouds.

An indication of air flow from the north, on the other hand, would tell you that far overhead is the eastern portion of an upper-air ridge. Just the opposite takes place here. The northward-then-southward hump in these winds encourages the clockwise flow and subsidence of lower-level air. This then encourages an area of high pressure and fair weather.

If cirrus clouds should be seen approaching *directly* from the west, take another look at Figure 7.1, then check the barometer.[4] If it reads high, you're likely sitting right on top of a ridge. If the barometer reads low, you're probably at the very bottom of a trough. (Remember in your cloud observations that the best way to judge the direction of cloud movement is to look directly overhead. By so doing you cut the distorting effects of perspective to a minimum, and your results will be that much more accurate).

Upper-air troughs don't *always* generate storms down here on the surface. But sometimes they do. And that just makes them all the more interesting to keep an eye on.

---

[4]For more information on the barometer and its indications, see Chapter 9.

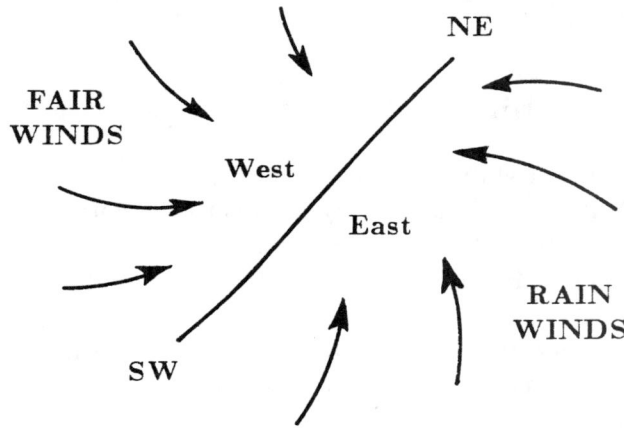

Figure 7.2: Fair-weather winds blow from the northwest quadrant. Southeast winds bring the rain.

Winds: Fair and Foul

> Draw out a line from Maine to LA,
> Wind to Miami, sun on its way.
> If the same line is drawn from LA to Maine,
> Wind to Seattle means likely rain.

The principle here is simply this: Winds from the southeast generally mean rain; winds from the northwest usually indicate fair weather (Figure 7.2). During a storm then, if the wind should shift such that it's no longer blowing out of the east-southeast-south quadrant, clearing weather shouldn't be too far away. But don't forget to watch the barometer along with the wind! By so doing, you can verify that what you're seeing in the wind is indeed what's taking place in the air over your head.

Wind Shifts

Wind that shifts like a clock henceforth,
 Will drop its rain on my friends to the north.
A wind that shifts the other way,
 Says, "My friends to the south are wet today."

To determine which way the wind is shifting—whether clockwise or counterclockwise—follow these steps:

- When you take a reading of the wind, look directly into it and notice which direction you're facing.

- Later, when you take your next reading, point yourself into the wind, and again notice which way you're facing.

- Now retrace your steps. Face the direction you did for your original reading, then slowly turn to face the way you did for your most recent reading.

- If your body turned clockwise ("like a clock"), the storm center is passing to the north. If, on the other hand, your body turned in a *counter*-clockwise direction, the storm is passing to the south.

Winds Shifts

If throughout a series of wind checks you seem to be finding no shifts at all, don't lose heart. A wind that isn't progressively shifting one way or the other but remains steady from one direction may indicate that a storm is approaching you straight on. If this is the case, the storm center is likely to pass directly over you—or very nearby.

> Wind shifting gradually east to north,
> Barometer low but rising;
> Predictors of a warming trend
> May soon be apologizing.

A wind out of the east or northeast likely means a storm is on its way and is going to pass somewhere to the south of you. The passage of the storm center to the south can be confirmed, of course, by noting which way your body shifts from one wind reading to another. If the storm then completes its passage to the south, you will be left a resident of its clear but cool northwest quadrant.[5] If summer is here, things may be just grand. The sun should warm things nicely. If instead, winter has made itself at home (as would be the case in the above rhyme), bitter cold could be yours for a while.[6]

---

[5] To keep your orientation straight in relation to the storm as it passes, you might find it helpful to refer to Figure 6.6 on page 55.

[6] Temperatures found in the northwest sector depend largely on the season. For a description of seasonal effects on temperature, refer to the rhyme on page 99 of Chapter 10.

> Wind from northwest,
>   Erratic at best.
> Wind from northeast,
>   More steady at least.
> From southeast, the gusts
>   Have almost ceased.
> And from the southwest,
>   With smooth air we're blest.

A wind from the northwest brings cold air down over warm land. In a situation like this, because air warmed by the surface of the earth will naturally try to push itself upward, unstable air results. Unstable air in turn causes strong updrafts (and thus downdrafts, too), bringing about erratic wind speed and direction at the surface. In addition to this, if a storm has just trekked by to the north, the passage of a cold front will be signaled by winds out of the northwest, and this will bring with it a blustery, though short-lived, squall.

Winds out of both the northeast and southeast are smoother than those from the northwest. But it's from the southwest that we get our smoothest air of all. Wind from this direction is typically found in the "lower right-hand corner" of a storm (see Figure 6.6 on page 55) where, as the storm approaches you, warm air overriding cold forms nice uniform layers in the atmosphere causing very little up-and-down motion, and thus very little gusting.

> Veering means the sky will be clearing.
> Backing means that rain won't be lacking.

Winds Shifts

To say the wind is "veering" is just another way of saying that the wind is shifting in a clockwise direction. Clockwise shifting, of course, means the center of the storm in question is passing to the north. Because of this, clearing can be looked for following the passage of a warm front, and again after the passage of a cold front.

To say the wind is "backing," however, means just the opposite. Here a *counter*clockwise wind shift is indicated, and with it, a storm passage to the south. That means the northern sector will pass your way. And that, as you'll remember from Chapter 6, spells r-a-i-n.

> Wind before cloud,
>   Long rain not allowed.
> Cloud before puff,
>   Long rain sure enough.

With the approach of a low-pressure system, the wind begins to shift and blow as it takes on the flow of the circular weather pattern. If this shifting about of the wind occurs *before* clouds begin to cover the sky, the storm should be short-lived. The signs here indicate that there is really very little moisture being lifted into the air and that the cloudy portion of the low probably won't be overhead for very long.

If, on the other hand, clouds come before the circular weather winds start to move counter to the upper-air flow, the approaching storm is obviously pushing large amounts of moist air into the atmosphere. Look out. The period of rain ahead could be a long one.

Chapter 8

Clouds

Identifying Clouds

> High-level clouds teach of upper-air winds;
> Middle clouds teach choreography.
> And what we learn from the low-level clouds
> Is a lesson in local geography.

Of all the characters that have ever performed upon the weather's stage, clouds are the undisputable favorites. Just think how drab life would be without them! They beckon us as if to say, "Hey, look up here, look up." And, oh, what rewards for doing so!

Clouds form and travel at three general levels: High, middle, and low. And clouds within each of these three levels have distinct messages to give us. Way, way up at the top are the high-level clouds keep-

Identifying Clouds

ing us informed as to how the westerlies[1] are currently flowing.

Traveling in between are those clouds which are not so high as to be caught by the high-level westerly winds, but not so low as to be detoured by the landscape. From these middle clouds we gain much valuable information about the weather-system winds as they swirl about the highs and the lows. It's through our observation of these clouds that we find most of our clues for making judgements concerning the coming and going of our daily weather.

Finally, there are those clouds which lie not too far above our heads. While the middle clouds are busily following the orders of the swirling weather-system winds, low-level clouds are in the process of being sidetracked (or perhaps even being formed) by the many humps and bumps of the earth's surface.

> Cirrus clouds like locks of hair.
> Stratus: level, straight.
> And wouldn't it seem quite logical
> That cumulus clouds accumulate?
>
> Cirro clouds like circus artists
> Fly so high to steal your heart,
> While alto clouds like female singers
> Sing the middle part.

The first part of this rhyme outlines three common

---

[1] For more information on the high-altitude westerly winds, see the first section of Chapter 4. See also the "Upper-Air Winds" section of Chapter 7 on page 69.

cloud categories, or *types* of clouds:

- *Cirrus* are those clouds notorious for impersonating the hairs of horses' tails. They are formed at very high altitudes and are composed entirely of ice crystals.

- *Stratus* clouds are those you see forming broad horizontal layers. They are associated with a stable atmosphere that isn't allowing much up-and-down movement.

- And then there are *cumulus*—the puffy clouds.

The second part of this rhyme deals with the common *prefixes* applied to cloud names to indicate the *altitude* at which the various clouds form. "Cirro" is a prefix for those clouds finding birth in the high altitudes. Puffy clouds, for instance, that form in the high levels of the atmosphere take on the name of *cirro*cumulus. From down here on the earth's surface, these often look like a myriad of tiny cotton balls, or like rippled sand. *Alto*cumulus, on the other hand, form a bit lower in the atmosphere and take on more the appearance of a flock of sheep. And just plain ol' cumulus are those clouds that puff and tower a bit closer to home. These are the ones that make favorites for trying to pick out familiar shapes—like dogs, ducks and dragons.

Combining a cloud-height prefix with the appropriate cloud-category name will supply you with the names of most of the clouds experienced in our daily

Identifying Clouds

weather. Following are the more common ones:

- **High Clouds** (20,000 feet or higher):
    - Cirrus: Horse-tail clouds.
    - Cirrostratus: Thin veil of ice crystals that may cause a halo around the sun or the moon.
    - Cirrocumulus: Tiny cotton balls, or rippled sand.

- **Middle Clouds** (between 7,000 and 20,000 feet):
    - Altostratus: Mid-level, light-gray veil that *almost* blocks the sun—but never completely.
    - Altocumulus: Lumpy flock of sheep.

- **Low Clouds** (those below 7,000 feet):
    - Cumulus: Puffy clouds.
    - Stratus: Straight, layered clouds.
    - Stratocumulus: Puffy clouds that aren't completely separated, but that aren't a nice smooth layer either.
    - Note: All of the low-level clouds are capable of completely blocking out the sun, and are often seen with dark gray undersides.

High clouds can sometimes be a bit fickle, playing tricks that may lead you to false conclusions about the weather's future. Low clouds, on the other hand, may be too localized to give you a good picture of the weather. It is from the middle clouds that you can read messages of greatest significance. They are close enough to affect your neighborhood, but they come from sources far enough away to have impact over a large area.

Cloud Tales

> How tall are the horses
> Whose tails fly so high?
> Five or six miles
> Away in the sky.

The moisture needed for these high-level cirrus clouds to form so high in the atmosphere very often has made its way from a storm that, while still far to the west, has pushed enough moisture into the air for some of it to be caught up by the westerlies[2] and swept on ahead to the east. Because of the much greater speed of the upper-air winds in relation to the weather winds below, cirrus clouds may forewarn of a storm even days before its actual arrival.

> When rain clouds make their way real low,
> They drench the earth from head to toe.
> When rain clouds pass a little higher,
> They leave the earth a little dryer.

Low clouds drop the heaviest rains—and for good reason. The rain drops falling from clouds creeping

---

[2]For more information on the upper-air westerly winds, see the first section of Chapter 4.

along close to the ground simply have less chance to evaporate and thereby to reduce in size before splashing on our umbrellas. Higher clouds, on the other hand, produce a finer, lighter rainfall.

> The lower they crouch,
> The bigger their pouch.

And remember, by the same principle, the *thicker* the clouds (meaning those whose bases are low, but whose tops extend a fair height into the atmosphere), the heavier will be their rainfall. Thus, if it is a thick cloud that crouches so low over your head, prepare for a soaking.

> The lower they get,
> The more likely wet.

Clouds generally lower and thicken with the approach of a weather-bearing low-pressure system. Thus, the lower the cloud deck becomes, the sooner the encroaching storm is likely to let down its rains.

> Thick enough to hide the sun.
> Thick enough to make you run—
> for shelter.

Only clouds found in the lower-levels are actually able to hold enough moisture to block out the orb of the sun. Cirrostratus clouds, made up of ice crystals in the high altitudes, will only "halo" the sun. Middle clouds will only veil it, never blocking it out completely. Low-level clouds, however, are able to build up enough moisture to make the sun disappear altogether. And when they've grown thick enough to do that, they're also thick enough to drop rain.

> Billowing tops, well defined;
> Always a chance of dropping some rain.
> Tops dividing, tops disperse ...
> They've decided to refrain.

Here's a good reason for keeping an eye on the *tops* of clouds. Sharp, well-defined "boiling" cloud tops indicate that updrafts are still pushing strongly. Moisture is still rising, still condensing, still a signpost of potential rain. When tops turn fuzzy and indistinct, they're signaling that upward flow has subsided and rain, if it fell at all, is a thing of the past for such clouds. The fuzziness is actually condensation that is dissolving, or being diluted, by mixing with drier surrounding air.

> Turrets and towers
> Will water the flowers.

If the clouds overhead start to look like medieval castles with their many turrets and towers, the signs are right for rain.

> Good flat bottoms and oval tops:
> Great for the picnic, bad for the crops.
> Randomly spaced with ragged feet:
> Great for the spinach you'd rather not eat.

Watch the undersides, too. A nice flat base is an indication that the cloud is still feeding and growing

Cloud Tales

from rising moisture. But keeping an eye on things could pay off. If the bases of the clouds start to turn ragged, rain could be dropping real soon.

> If clouds march forth
> In rank and file,
> Dropping much rain
> Simply isn't their style.

The rank-and-file, all-in-a-line style of cloud formation is generally just fair-weather moisture doing its thing. Some isolated drops may fall here and there, but all in all, rain shouldn't be a cause of much worry. These "cloud streets," as they're often called, are formed when wind blows cool air over warmer, moist ground. Bubbles of moisture build at the surface then break loose into the atmosphere. These bubbles then create "puffs" of cloud which tend to form in lines as they are released one after another and are blown along by the wind.

> High, small cumulus
> Out of the west:
> Their presence is fair weather
> Being expressed.

Fair-weather cumulus clouds are an especially common sight after the passage of a cold front. Here, in the cooler northwest quadrant of a passing storm, cold air moving over warm earth causes bubbles of warm, moist air to break loose from the surface and rise into the air forming clouds. It just goes to show that not every cloud means rain. While clouds of this

Thunder Storms

> How far the storm?
> How else to derive?
> But count out the monkeys
> And divide by five.

It takes about a second to say the words "one monkey." And because sound travels about 1,100 feet per second, a count of *five* monkeys from lightning strike to thunder clap, tells you the storm generating the thunder is about one mile away. Just remember: One monkey is equal to about one-fifth of a mile, so all you have to do is "count out the monkeys, and divide by five."

It's interesting to note that thunder can't be heard when the originating storm is over 25 miles away. And thunder is usually heard only when storms are within eight miles. This explains lightning that is often seen with no accompanying thunder. Light, which travels so fast and far reaches your eyes quickly while sound, which travels so much slower, often never reaches your ears at all.

> Lightning flash from west-northwest,
> You'll likely be its honored guest.
> Lightning from another way,
> Says, "Sorry, perhaps another day."

Thunder storms generally travel an approximate northwest-to-southeast path. Because of this, if you see lightning coming from anywhere other than out of the northwest, the storm will likely pass you by. And take note, too, that lightning strikes generally occur at different levels, depending on the season. Hot-weather thunderstorms will release their lightning at fairly low levels. Cold weather storms, on the other hand, are more prone to pass their lightning from cloud to cloud.

>   Once the anvil
>   Shows it head,
>   The storm's a beast,
>   But'll soon be dead.

A thunderstorm is essentially a towering cloud whose upward push of moist air has reached extremes. As moist air continues to rise, the cloud top pushes higher and higher into the atmosphere until finally it emerges into the stream of the upper-air winds. The cloud top is then caught up by these winds and is dragged along by their flow. And once the top finally starts to be dragged from a thunderhead, such tremendous height indicates to us that it has reached its most powerful state. But, in addition to this, the self-perpetuating power that has fueled the cloud is now being sucked out of it, and its ferocity is quickly being drawn to an end.

More Cloud Tales

>   When puffy clouds look smaller
>   At sunset than at noon,
>   The signs look right for a stroll tonight
>   Beneath a silvery moon.

The reference here is to puffy cumulus clouds which are very often the result of temperature irregularities caused by nothing more than the heat of the day. Therefore, if the clouds that appeared during the day should begin to shrink with the coming of evening, you can rest assured that nothing of any longer-range consequence was meant by them.

> When clouds from west or north
>   Let big blue spots of sky show forth,
> The spots will grow larger to show more blue,
>   And clear skies will last for a day or two.

When a storm is just completing its passage, a signal that rarely fails to indicate coming fair weather is the appearance of ever-enlarging blue patches to the northwest. They are quite literally the "breaking" of a storm whose clouds have grown weary of hiding the sun.

> The longer the fair-weather stretch,
> The less certain picture high-level clouds sketch.

The formation of clouds in the upper levels of the atmosphere are very often an early warning of a storm yet unseen. But, like most things of the weather— not always. If the skies have been fair for a good long time, the weather picture painted by these high-level clouds becomes less and less reliable. They may appear, and then disappear again, with nothing to follow but more good weather. Under such circumstances, the key to making these clouds useful to you as a predictor of the weather's changes is to use your

More Cloud Tales

barometer to verify their indications. If the pressure stays steady, go ahead and pack the car—the camping trip is on. If, on the other hand, the barometer should begin its plunge, your observations could very well pay off by saving you from vacationing under an atmospheric blanket of gray.

> Contrail streamer in the sky,
> Showing me where airplanes fly.
> If your tail should melt away,
> Tomorrow'll be another fine day.
> But if your tail should stick around,
> Tomorrow rain may wet the ground.

Condensation trails, better known simply as contrails, are clouds created by the moisture being expelled into the atmosphere by the hot exhaust gases of jet aircraft. Like cirrus clouds, they are made up of ice crystals and occur only at high altitudes.

Because they are inadvertently placed in the sky by an airplane just happening by, contrails can sometimes give us clues to weather conditions otherwise unobservable. If a contrail should dissipate quickly, for instance, we have a sign that the air way up there is fairly dry (because it so easily dissolves the moisture), and that fair weather is still with us for a while. On the other hand, a persistent contrail indicates moist air. And if there's moisture floating around up there, chances are that it may be lowering soon. This is particularly true if the appearance of cirrus clouds and a falling barometer confirm your suspicions.

> If God should will it,
> The sky will spill it.

Some clouds drop their rain, and some clouds don't. Some drop it when they shouldn't, and others don't when they clearly should. So when all is said and done, an important principle for any weather-watcher to keep tucked in his arsenal of forecaster's ammunition is this: "When it rains it rains, and when it shines it shines." And few there are who can argue with that.

Chapter 9

Barometric Pressure

The Barometer in General

> How to know what's happening o'rhead?
> That is why the barometer is read.

Unlike some fish that swim among the waves, it just happens to be our lot in life to live out our days at the *bottom* of the great atmospheric ocean. We have constantly above our heads a great mass of air bearing down with its weight upon all that is below. Because we've grown accustomed to its load, subtle fluctuations for the most part escape our notice.

And that's where the barometer comes to our aid. The barometer is our scale upon which we are able to determine just how much of the atmosphere happens to be sitting overhead at any given time. If the mound of air should roll up particularly high, then "high" would be the reading on our barometric scale. On the other hand, should there be a valley passing over our heads, our scale of course would read "low." This simple, inexpensive instrument is the handiest tool we have to help us in determining, for instance, whether that halo we see enclosing the moon is really the early warning of an approaching low, or whether it is just a wisp of harmless high-level moisture. If our observations confirm that the halo is indeed being accompanied by a dip in the barometer, we had best take note. As you can see, the barometer adds that little something to our analysis that otherwise would remain unfelt and perhaps unseen.

As a rule, a *drop* in barometric pressure means cloudy weather. A *rise*, on the other hand, means fair. And to keep that straight, just remember this: When the barometer falls, so does the rain.

Reading the Ups and Downs

> Pressure's high,
> Not a cloud in the sky.

The chunk of atmosphere contained in a high-pressure weather system is characterized by clear skies. And there's good reason. A high-pressure area is a large, slowly sinking mass of air. As a result of this sinking, the mass is being compressed. Compression, in turn, warms the air, and before you know it, moisture has been evaporated and clear sky is the norm.

> Pressure's high,
> So am I.

High pressure is something most people look forward to. (High-pressure *weather*, that is). A prediction that a high-pressure system is moving into the area is somewhat like a promise of sunshine. And most people like to hear promises like that.

> Drop the pressure a little too low,
> Ol' Rover'll feel it head to toe.

Air pressure does affect the nerves. And low pressure does it most adversely. In humans, forgotten sores begin to ache. Corns and bunions, forgotten in fair weather, make themselves known again with the lower pressure of foul weather. And there's no end to the symptoms attributable to fluctuations in barometric pressure. Keep your weather eye on dogs and other pets (particularly those who have already seen the better portions of their lives). They too, just like humans, are likely to show visible signs of suffering from the well known "ache of the weather."

Low pressure typically means air is on the rise. As this air makes its way up into the atmosphere it expands, causing the air to cool. Cooling in turn brings about condensation, and condensation brings us clouds. And that is exactly what low-pressure areas are best known for—clouds, and of course, rain.

>Pressure's finally on the rise,
>Wind is from the west-northwest.
>The cold could make you agonize,
>Though skies will be their sunniest.

The indication here is that a storm has finally made its passage on to the east, leaving us in its cold northwest sector. The rise of the barometer signals the end of the low and the start of the high. And the rise in pressure, of course, will bring clear skies to our area.[1]

>Pressure falling, winds southeast,
>Rain for a couple of days at least.
>Winds northwest, pressure steady,
>For a couple days of sun be ready.

A dropping barometer accompanied by wind from the east is like having the weather itself speak the words, "Rain is on its way." If the pressure has steadied and the wind is now from the northwest, its message is, "Clear skies are going to be around for awhile." And because the average passage time for our weather systems is about 48 hours (*average*, remember), whatever the barometer and wind should happen to say is heading our direction, we can plan on a couple days of it once it gets here.

---

[1] Temperatures found in the northwest sector depend largely on the season. For a description of seasonal effects on temperature, refer to the rhyme on page 99 of Chapter 10.

Reading the Ups and Downs

> Wind up, barometer down,
> Unpleasant weather is coming to town.
> The higher the breeze and slower the fall,
> The tougher the storm 'twill be for all.

A rising wind and a falling barometer is just like having a two-headed town crier shouting, "Here comes a storm." Add some strength to the winds and a nice slow fall to the barometer and you can be pretty sure it's not just any old storm that's happening your way. You had best get ready for a blow.

Chapter 10

Temperature

Stable and Unstable Air

> Where hot meets cold
> Beware and behold,
> That is the spot
> To witness a lot.

Temperature contrasts are what make the weather. And the greater the contrasts, the better the performance offered by the events of the atmosphere. This is true particularly along the frontal lines of the wave cyclone[1] where frigid air from the north is brought face-to-face with warm air from southern regions. The reason behind all the violence is simply this: Where hot air and cold air meet, hot air pushes

---

[1] For more information on the wave cyclone, see the first section of Chapter 6.

Stable and Unstable Air

upward. And when hot air pushes upward, clouds form, winds blow, rains fall and, on occasion, lightning strikes and thunder crashes.

> When clouds form in layers,
> The air is asleep.
> But when 'tis awakened,
> The clouds form a heap.

When we speak of the "stability" of the atmosphere, we are referring to the amount up-and-down motion taking place in the air. If you get warm air sitting on top of cool air, the warm air on top simply stays there. Lack of any significant temperature differences in the air farther up means there's nothing to encourage lifting. Down below, cooler air sits contentedly like a mass of jello, having no intentions of rising. With so little exchange between the two layered masses of air, winds are light and skies are either clear or, if skies are overcast, the clouds form a nice even blanket, unbroken by any up-and-down movement of air. The latter often results from the moisture in the warmer air above condensing into clouds where it contacts the layer of cooler air below.

> When warm atop cold
> Concludes its excursion,
> The result of it all
> Is called an inversion.

In the typical atmosphere, you'll find air getting progressively cooler the higher you go. To have a warm layer of air on *top* of a cooler layer is therefore referred to as an "inversion." An inversion may come about in a number of ways:

- One cause of an inversion is to have the surface of the earth so cooled by radiating its heat into space that the earth in turn cools the air closest to it. This creates a surface layer of air that's cooler than the air farther overhead.

- Another cause of an inversion is to have a warm mass of air override a cooler mass.

- And yet another cause is to have a mass of sinking air—as in a high-pressure weather system[2]—warmed by compression as it descends and eventually stops to sit quietly atop a mass of slightly cooler air below.

In all these cases, the wind (if present at all) is steady. The sky is either clear, or any moisture present is represented in the form of layered clouds. Remember, an inversion represents stable air. And stable conditions tend to last awhile.

> Hot above, cool below,
> Clouds have little chance to grow.
> Hot below and cool above,
> Upward flow is push and shove.

Towering clouds are the result of up-and-down motion in the atmosphere. And up-and-down motion in the atmosphere is largely a result of differing

---

[2] For more information of the formation of high-pressure systems and their role in our weather, see the first section of Chapter 2.

temperatures. When warm air sits above cooler air, vertical movement is slight. Put the warm air below and the cool air above, however, and the story has an entirely different theme. The warm, moist air below is now just itching to be on the rise, and the cool air above is just urging it to come up. Alternating pockets of blue sky and billowing clouds typify the instability found in such air.

Unseasonable Temperatures

When temperatures outside
   Seem a bit unseasonable,
The reasons for it happening
   Usually aren't so unreasonable.

In summer, low pressure
   And rain make it cool,
While high pressure's sun
   Makes you head for the pool.

In winter, low pressure
   And rain make it warm—
A cloudy warm blanket
   In spite of the storm.

But winter high pressure
   And skies without trace,
Make all the earth's warmth
   Disappear into space.

   Unseasonably high or low temperatures occur under generally predictable conditions. In summer, low pressure causes a cloudy blanket to block the sun's rays, and the temperature drops right along with the rain. Summer high pressure, on the other hand, allows the sun to constantly bathe the earth, making temperatures unusually high.

In winter, the tables are turned. Coldest temperatures are found where the sky is clear. Instead of the sun warming things up as it did in summer, now the heat is being radiated back into space faster than it can be replenished. The result is unusually cold weather. Higher-than-usual winter temperatures come when a cloudy blanket is present to keep the earth's heat from escaping into space. This happens most often when a low-pressure system is passing over, blanketing the area in overcast.

Wind Chill

> Every five of blow,
> Drops it five degrees below.

Wind chill isn't something that can be read from a thermometer. Instead, it is a phenomenon we *feel*. If you were to step outdoors into air your thermometer said was 40 degrees Fahrenheit only to find that a 15 mile-per-hour breeze was blowing, your body would ignore what the thermometer had told it, feeling instead as if it was actually a very chilly 25 degrees!

When the wind blows past you, warmth is drawn from your body much faster than it would if the air was standing still. The result is called "wind chill,"

Wind Chill

and can be roughly calculated in your head by following these steps:

1. Read the temperature from the thermometer.

2. Make an estimate of the wind speed.[3]

3. Subtract the wind's speed (in miles per hour) from the temperature (in degrees Fahrenheit). The resulting temperature is now corrected for wind chill.

Here's an example: If you read a temperature of 30 degrees on the thermometer and estimate a wind speed of ten miles per hour, the wind-chill factor will drop the temperature to 20 degrees as far as your body is concerned. Wind chill is a nice thing to be able to estimate, and there are times when being able to make such an estimate can be downright important to your comfort—and perhaps even to your life!

---

[3] For help in determining the speed of the wind, refer to the rhyme on page 67 of Chapter 7.

Chapter 11

Humidity and the Dew Point

Understanding Humidity, Dew, and the Dew Point

> How else to say it?
> How else to construe?
> 'Cept, "The dew point is
> The point of dew."

Dew is much more than just the beady droplets that form on your front lawn. Dew includes any condensation of moisture on any cool surface—a list of which includes grass, glass, roses, hoses, cars, spars, bicycles, tricycles, and believe it or not, even tiny particles floating about in the air called "cloud-condensation nuclei." So you see, dew is not something peculiar only to the ground. Dew forms on many things—

Understanding the Dew Point

both here on the ground, and in the air, too. And clouds are one of the many results.

Humidity is dampness—the unseen moisture of the atmosphere. And while dew is that exact same moisture condensed into a liquid form that we can see and feel, the *point* of dew—or the "dew point"—is something else entirely. When we make reference to the dew point, we're actually referring to the *temperature* at which moisture in the air will condense into dew. So remember: Humidity is the unseen moisture. Dew is that same moisture, but in liquid form. And the dew point is a temperature.

> As temperature's love waxes warm and cool,
> The dew point remains ever true.
> And when e'er the two should meet again,
> In the air will be fog, and on the ground, dew.

Temperature of the air fluctuates up and down for any number of reasons. The dew point, however, is the point to which the temperature must drop before moisture in the air will condense into visible form. Therefore, when the temperature of the air finally lowers to the temperature of the dew point, we find clouds, fog, or dew.

> So long as temperature and dew point keep
> Their distance, and so remain,
> So long will be the time during which
> The skies withhold their rain.

In order for there to be rain, there first have to be clouds. And clouds are formed by condensation of moisture in the atmosphere. It's easy enough to see that if the temperature of the air never drops low enough to reach the dew point (and thereby get the condensation process into motion), the skies have no other choice but to remain clear.

Estimating the Dew Point

> Wet your finger, moisten it well,
> Hold it high above your head.
> The colder your finger feels up there,
> The greater the temperature –
> dew point spread.

With a low dew point, air is still able to hold a lot more moisture, and thus the moisture on your finger is drawn easily and quickly into the atmosphere. This, in turn, causes your finger to feel cool—perhaps even cold. A higher dew point, on the other hand, results in just the opposite: Evaporation is slow, and the sensation of coolness is slight if felt at all.

Note that this test will work best after some experience is gained in its use. Only through practice and comparison will your judgement become refined enough to tell you how your finger *really* feels in relation to the current state of the dew point.

Estimating the Dew Point

> If you step outside
>   And your hair goes limp,
> The dew point is likely
>   As high as a blimp.
>
> Or, if you're like some other folks,
>   Whose hair would rather frizzle;
> The dew point is likely just as high,
>   And may forewarn of drizzle.

Limp or frizzy hair (whichever the case may be) is a common sign of high humidity. And high humidity means a high dew point, indicating that the air temperature doesn't have to drop very far before clouds will begin to form. The reverse is also true. If you step outside and your hair is lively—blowing freely in even a slight breeze—the dew point is likely good and low.

> Doors and windows like to stick
> When moisture in the air gets thick.

When an air mass is passing through that happens to be a bit on the moist side, things start to happen around the house. Wood, in particular, likes to swell when the air gets damp, and close fitting items like doors and windows can become a bit tough to open and shut. Keeping an eye on such things can offer clues to what lies ahead in the weather. Connecting up a sticking door with a slowly falling barometer, for instance, just might provide your first hint of an approaching storm.

> Murky air, hazy sky,
>   The dew point must be pretty high.
> Platinum moon and stars so bright,
>   The dew point must be low tonight.

Often times you can "see" and "feel" moisture in the air even before it condenses into clouds or fog. Your moisture sensors—your sight and your general sense of feel for mugginess in the air—are one way of estimating the dew point. If the air is "loaded," you know it won't take much drop in temperature to bring about condensation which is *the* sign that the dew point has been reached.

> The height of the clouds
>   Is a visible clue,
> Of how high or low
>   Is the point of dew.

The typical condition of the atmosphere is that the higher you go, the cooler the air gets. If the dew point is low, air will have to rise quite high into the atmosphere before it reaches air cool enough to condense its moisture. On the other hand, air with a *high* dew point will not have to rise very far before it finds temperatures cool enough to turn its moisture into clouds.

Tales of Dew, Fog, Thunder and Frost

> No fog nor dew on a cloudy morn?
> A cloudy quilt kept the earth too warm.

For fog or dew to form, the temperature of the earth's surface must first become low enough to reach the dew point. And for it to drop this low in temperature generally requires that the night sky be nice and clear to allow the earth to radiate its heat off into space. Once this has been accomplished, air close to the ground can be cooled enough by the earth to bring it to the dew point. The most common hinderance to the formation of dew or fog, therefore, is some sort of cloud cover that will keep the earth's heat from escaping into space.

Fog generally comes in two forms. The first form appears when the surface of the earth is cooled by radiating its heat into space. This type is commonly referred to as radiation fog. The other form is advection fog, produced by warm moist air being blown into contact with the cold earth. This happens often when air from the sea is blown onto land, covering the coastline in a dense fog.

> A dewless morn
> Of rain doth forewarn.

After a stretch of good weather, a dewless morning could well be the first sign of change. Sometimes even a thin, almost imperceptible layer of high-level moisture may be enough to prevent dew from forming. And the thin, high clouds that prevent the dew could well be the first criers of a storm on its way.

> If morning starts you off to work
>   With squeegee in your hand,
> Then plan to eat your lunch outdoors
>   Or take some sick leave you hadn't planned.

A squeegee in hand spells almost certain sunshine. During the night, the earth has been busily releasing its heat into space, and the current state of the dew point is expressed all over your car. Without clear skies, the dew wouldn't have formed.

> When morning dew glistens,
>   Remove all suspicions,
> The sun, it will rise
>   To light up blue skies.

Dew is a sure sign of fair weather. But very often, the dew you find on the ground will be accompanied by fog in the air. The two are often a joint product of cloudless night skies. If you're not careful, though, you may be tricked into thinking the fog is really a cloudy sky and that the dew is just a mistake. But you can be sure, if there's dew on the ground, those aren't really clouds overhead. Real-live clouds prevent dew from forming. Fog, on the other hand, is actually a *result* of clear skies, and unless it has formed a particularly thick layer, the fog will dissipate quickly once the sun is up.

Dew, Fog, Thunder and Frost

> If temperature shows
> A steady climb,
> The fog will lift
> In a matter of time.

Once the sun has risen and gets to work warming things up, the earth starts to soak a bit of it in. Before you know it, the ground is warm enough to start evaporating some of the fog nearest its surface. As this goes on, the fog continues to be evaporated from the bottom up, giving the impression that it is indeed "lifting." So, as long as the temperature continues to climb, the fog will eventually dissipate.

> Morning dew,
> Skies of blue.

If you've risen early on a Saturday morning and suddenly find yourself having second thoughts about packing the picnic basket, just step outside and brush your hand over the lawn. If it's wet with dew, get on with the packing. But if you discover the lawn to be dry, look for other signs to help you discover what sort of mischief the atmosphere might be up to.

> Sultry morning,
> Fair warning;
> Not to blunder
> 'Bout afternoon thunder.

When the air feels especially hot and muggy, it serves well as an early-warning indicator that afternoon thunderstorms are not to go without consideration. Dry air is comprised primarily of nitrogen, a very heavy gas. When saturated with water, however, the same volume of air consists mostly of hydrogen dioxide, among the lightest of gases. Therefore, once the sun goes to work further lightening the damp air with its heat, the ingredients are suddenly right for the strong upward movements of moisture that are ideal for the breeding of thunderstorms.

> Cold, calm, and clear,
> Jack Frost is passing near.

Clear skies encourage the formation of dew. *Cold* clear skies are the forerunners of frost—dew in its frozen form. Add "calm" to the cold and the clear, and you have a combination that frost can rarely resist.

> When your tomato plant cries,
> "It's freezing out here!"
> The weather is probably
> Cold, calm, and clear.

On calm, windless nights, the atmosphere tends to sort itself into layers, with the coldest air settling to the bottom. And where there's cold, calm air, frost is likely. Even a little wind may be just enough to prevent this layering, and thereby prevent frost from forming. The time to be especially on the lookout for frost is with the passing of the cold front when skies clear, temperatures plunge, and under an expanse of high pressure, nights become still.

Clouds (or the lack of them) are important, too. Even very high clouds are able to prevent temperatures from falling as low as they might otherwise. So, if you're concerned about the possibilities of frost, watch the trend of the clouds. If toward evening they seem to be thickening, don't worry, frost is unlikely. If, on the other hand, the clouds appear to be breaking up, watch out.

Chapter 12

Other Weather Phenomena

Smell

If with your nose you "smell" the day,
Stormy weather's on its way.

Have you ever thought you could smell a coming change in the weather? Well, "sniffing" the weather is indeed a valid part of your weather observation. High pressure that accompanies fair weather tends to keep scents and odors dormant. When a low-pressure system replaces the high, these scents are ever so gently released, enabling us to "smell" a storm on its way.

Sight and Sound

>Whistle soft,
>>Whistle loud,
>
>Skies clear,
>>Skies cloud.
>
>Mountain far,
>>Mountain near,
>
>Clouds will make it
>>Disappear.

Our weather theatre, being the wondrous structure that it is, has been built to accommodate a variety of acoustical qualities. Sometimes the ceiling is infinitely high and sounds produced here on earth start a journey into endless space, never to echo again in our ears. The result is that distant noises are faint, if perceptible at all. Often enough, though, a temporary ceiling of clouds is lowered into place to form an atmospheric sounding board which makes audible even those sounds normally too distant to hear.

Similar things happen with sight. Distant mountains, normally obscured by haze, will suddenly appear as though you could reach out and touch them. Every detail is vivid, and landmarks once forgotten become clear again. This often occurs just before a storm.

>Like the big empty sound
>Of a big empty room,
>So echo the sounds
>>Under overcast gloom.

As a storm makes its approach, many things begin to change. Things smell different. Things look different. And, things sound different. Indeed, the weather is a show for all the senses!

Halos

> Ring around the moon,
>   Rain by noon.
> Ring around the sun,
>   Rain before the night is done.

A ring—or halo—formed around the sun or moon is a signal that high-level moisture is overhead. The moisture being proclaimed here is composed entirely of ice crystals which refract[1] the light of the sun or moon in such a way that the resulting optical effect is a halo. Halos are often the predecessors to the horsetail cirrus clouds, which in turn are often signposts to an approaching low.[2] A halo, then, could well be your very first visible sign that a storm is building yonder somewhere to the west.

> The bigger the ring 'round sun or moon,
> Less likely "maybe" and more likely "soon."

As the high-level layer of moisture producing the halo around the sun or moon gets progressively lower,

---

[1] Refraction is the process by which rays of light are deflected, causing them to change direction as they pass through a medium—in this case, ice crystals.

[2] Refer to Figure 6.5 on page 53 for help in visualizing this progression of events.

its lowering also causes the halo to appear progressively larger. And lowering moisture is a likely sign that a weather change is on its way. But check the barometer before cancelling the picnic! While that high-level moisture causing the halo is very likely related to an approaching storm, things don't *always* work out so. Beware also that halos appearing in winter are much less reliable indicators of clouds to come because, with colder temperatures, high-level ice clouds are more prevalent and thus less likely to be directly related to an approaching weather system.

The Moon and Stars

> Sharp moon, stars bright,
>   Go to sleep in peace tonight.
> Dull moon, stars pale,
>   Dreams of sun will not prevail.

Sharp, bright stars and moon indicate dry air, which spells clear skies for a while at least. High-level moisture has a tendency, on the other hand, to obscure the moon and stars, making them hazy and indistinct.

> Pale moon,
>   Rain soon.
> Face white,
>   Rain chance slight.

The color of the moon varies with the amount of moisture present in the atmosphere overhead. A stark white moon, with a clean, sharp outline means very little moisture is present and clear weather should continue. If its outline turns dull, however, and its face a bit pale, start looking around for other signs of approaching stormy weather.

> Twinkle, twinkle little star,
> Whispering that rain's not far.

When you look at objects through the rising heat of a campfire, images flicker and distort. That's what happens when warm and cool air mingle. A similar interaction of warm and cool air occurs high in the atmosphere when warm air overrides cool air with the approach of a storm's warm front.[3] Twinkling stars are the result, giving early indication of an approaching storm. So when the stars twinkle, keep an eye on things to the west, and check the barometer frequently.

Sunsets

> On a summer evening facing west,
>   If clearly you see the sun enter her nest,
> Then don't you worry, and don't you fret
>   'Bout waking tomorrow and finding things wet.

Here in the northern hemisphere, our weather has gotten in the habit of making its approach on us from the west. If at sunset, therefore, you have the privilege of seeing the sun disappear over a clear horizon

---
[3] Refer to Figure 6.5 on page 53 and Figure 6.6 on page 55.

rather than over a bank of clouds, it's likely there's yet enough clear sky out there to make tomorrow another nice day.

> Red sky at night,
> Dry dust scatters light.
> Sky not so red,
> Moist air's there instead.

Though the beautiful red of a sunset is a pleasant color, mariners of old had much better reasons for finding delight in a red evening than in its beauty alone. The color of the sky is constantly changing according to the types and sizes of the tiny particles that happen to be floating around in the atmosphere. And it is the *size* of these particles that provide us with visual clues to aid us in determining whether wet or dry air is coming our way.

Particles of dry dust are larger than the typical particles on which moisture likes to condense. And it happens that dry dust in the atmosphere scatters the sun's spectrum in such a way that red becomes more prominent to the viewer. Thus, when the sun makes the evening sky glow red, what we are actually seeing is the colorful result of the sun shining at us through a *dry* mass of air to the west. And because our weather comes primarily from the west, when the sky glows red, dry weather is sitting on our horizon.

A pale sky, on the other hand, is signaling that there is moisture in the air. When moist air is to the west, the sun shining low through the atmosphere appears more gray or yellowish. The particles on which moisture typically likes to condense are a good bit smaller than the dust of the red sunset. As a result, more of the blue component of the sun's spectrum is allowed to show. The indication is that, if the air to the west has turned gray or yellowish in appearance, moist air is on the western horizon, and it won't be sitting there for long. So prepare for rain.

Remember, the indicator here is the sky, not the sun. Rather than analyzing the sun itself, you want to be checking the color of the sky *surrounding* the sun. And remember, too, that while a sun*rise* can be just as beautiful as a sunset, sunrises occur in the east. What we see in them are phenomena taking place in air that has already passed our way. So enjoy the sunrises for signs of what has already passed, but look to the sunset for what lies ahead—to the west.

Rainbows and Bees

> Rainbow in the morning,
> Picnicker take warning.
> Rainbow in the afternoon,
> An evening stroll is opportune.

The principle here is an easy one: Wherever you see a rainbow appear, you can be sure you're looking at a good chunk of moist air. Because weather travels as it does in a general west-to-east direction, if the rainbow appears to the west, rain may very well be dropping in your neighborhood soon. However, if the rainbow should make its appearance in the east, chances are things will be clearing for a while.

> If a bee's in the flower,
> There won't be a shower.

If after all this you're still procrastinating over packing the picnic basket, put your worries aside for a moment and go on out to the flower garden. Now, take a good look for any bees that might be buzzing around. If you find they're out smelling the flowers along with you, then get on with the packing! As long as they're anywhere but in their hives, rain won't be falling unexpectedly.

Index

Words from rhymes appear in italics

A

acclaim, of weather predictor 4
accumulate, cumulus clouds 79
ache, of the weather 93
acoustics, cloud ceiling 113
advection fog, described 107
adventure, wind stories 65
agonize, rising barometer, cold 94
air,
    weight of damp air 110
    weight of dry air 109
air mass, characteristics 38
air mass,
    defined 37
    air-mass weather 38
airplanes, contrails 89
allowed, long rain 77
altocumulus clouds, described 80, 81
altostratus clouds, described 81
always right, Coriolis force 25
anticyclone, defined 48
anvil, thunder storm 87
apologizing, predictors of warming 75
April, location of the westerlies 35
arguing, differences in temperature 11
asleep, stable air 97
atmosphere, typical, stability 97
atmospheric pressure, rising air 15
average, weather-system speed 44
average weather, climate 40
awakened, unstable air 97
axis, tilt of the globe 33

B

backing, wind 76
bacon, dry summer 10

Words from rhymes appear in *italics*

barometer,
    atmospheric scale 91
    general observation 5
    ridge, trough 72
    rising and falling air 15
    monitoring high-level clouds 88, 89
beast, thunder storm 87
bee, rain showers 119
behold, where hot meets cold 96
belly, earth's belt of warmth 33
below, wind chill 100
belt, earth's belt of warmth 33
bet, facing west 30
beware, winter storm 61
billowing tops, cloud tops 84
blame, of weather predictor 4
blanket,
    unseasonable winter temperatures 99
    warm front 52
blimp, estimating the dew point 105
blow, wind chill 100
blue skies, dew 108
blue spots, breaking of the storm 88
blunder, thunder 109
bolt, tighten, high pressure 29
bone, winter storm, wind chill 61
bottoms, clouds, rain 84

bowling ball, cold front 56
bread, wet spring 10
brick wall, cold front 56

C

Cancer, Tropic of 36
Capricorn, Tropic of 36
cart, friction and Coriolis 42
center, locating storm's center 51
chart, weather 16
chill, wind chill 100, 101
chill, winter storm 61
choreography, middle clouds 78
circle, weather symmetry 45
circular winds, caused by friction 42
circus artists, cirro clouds 79
cirrocumulus clouds, described 80, 81
cirrostratus clouds, described 81
cirrus clouds,
    described 80, 81
    directly from the west 72
    from the north, ridge 72
    from the south, trough 71
    height of 82
    warm front 53
    wind indicators, westerlies 69

Words from rhymes appear in italics 123

climate, defined 40
clock, wind shifts 74
clockwise,
    anticyclone swirl 48
    determining wind shift 74
    upper-air ridge 72
    veering wind 77
    wind around a high 29
cloud street, described 85
clouds, frontal passage 57
clouds,
    categories, prefixes 80
    formation, rising moisture 18
    general observation 6
    high-level, described 81
    high-level, reliability 88
    high-level, what they teach 78
    judging speed of movement 68
    low-level, described 81
    low-level, rain 82, 83
    low-level, what they teach 79
    middle-level, described 81
    middle-level, what they teach 79
clue, dew point, height of clouds 106
cold front,
    characteristics 56
    formation, 51

color, of sunset 117, 118
compresses, sinking air 18
compression, high pressure 92
construe, dew point 102
contour map, wind flow 16
contrail, indications 89
convergence,
    lifting force 21
    upper-air ridge 71
Coriolis force, described 25
corn, dry summer 10
cotton, south unseasonable 10
counterclockwise,
    backing wind 77
    cyclone swirl 48
    determining wind shift 74
    upper-air trough 72
    wind around a low 29
criers, warm front 54
crouch, rain clouds 83
crow's, locating storm's center 51
cumulus clouds,
    described 80, 81
    fair-weather 85
    shrinking at evening 87
currents, sea of air 2
cycle,
    of the winds 16, 22
    of water 22

cyclone,
    defined 48
    eddy cyclone 62
    wave cyclone, winter 61

D

darling, Coriolis force 24
dates, start of seasons 36
dead, thunder storm 87
detour, Coriolis force 26
dew, blue skies 108
dew,
    defined 102
    formation 107
dew point,
    defined 103
    estimating with moist finger 104
dewless, morn 107
divergence, upper-air trough 71
door, location of westerlies 34
doors, sticking, humidity 105
drama, sky 1
dreams of sun, dull moon, stars pale 115
drench, cloud height, rain 82
dust, winds to trust 66
dust, color of sunset 117

E

earth, tilt of 33
east, wind from 94

echo, sound, approaching storm 113
eddy, cyclone 62
equator, warm air rising 13
equinox,
    autumn 37
    spring 36
erratic, wind from northwest 76
errors, weather signs 8
Eskimo, winter storm 61
excursion, inversion 97
expands, rising air 18
expansion, low pressure 93
expressed, fair-weather clouds 85

F

facing west, upper-air westerlies 30
fall, starting date 37
feast, sky 1
female singers, alto clouds 79
ferris wheel, wind flow 14
finger, estimating dew point 104
fingers, to depict a wave cyclone 47
fish, sea of air 2
five-hundred,
    speed of westerly flow 32
    weather-system flow 44
flags, estimating wind speed 67

Words from rhymes appear in italics

flight, warm air rises 12
floods, winter storm 61
Florida, frost 10
flower, rain, bees 119
flowers, rain clouds 84
fog,
    advection fog 107
    formation 107
    lifting of 109
    radiation fog 107
forces, lifting 20
friction, effect on Coriolis 42
fringe, warm-front blanket 52
frizzle, hair, estimating dew point 105
front,
    cold, characteristics 56
    cold, wave cyclone 51
    occluded 58
    polar 49
    warm, characteristics 52
    warm, wave cyclone 51
frontal passage, identifying 57
frontal zones, lifting force 20
frost, conditions for 110

G

geography, low-level clouds 78
get it, facing west 30
glimpse, westerly air flow 32
glistens, morning dew 108
globe, tilt of 33
God, rain 90
gone by, weather symmetry 46
guess, weather signs 8
guest, thunder storm 86
gust, winds to trust 66
gusts, related wind direction 76

H

hair, estimating the dew point 105
halo,
    around moon and sun 114
    warm front 53
halves, weather symmetry 45
ham, dry summer 10
hand width, judging cloud movement 68
hands, to depict a wave cyclone 47
hazy sky, high dew point 106
heap, unstable air 97
heart, cold front 57
heat, lifting force 20
height, of clouds, dew point 106
hemisphere, see northern hemisphere
high clouds, types of 81

Words from rhymes appear in italics

high pressure,
    direction of flow around a high 29
    wind flow high to low 16
    fair weather 92
historical sequence 6
horse, friction and Coriolis 42
horse tails, upper-air ridge, trough 71
horse tails,
    warm front 53
    wind indicators, westerlies 69
horses, height of cirrus clouds 82
humidity, frontal passage 57
humidity,
    defined 103
    general observation 6
hup, location of westerlies 34
hydrogen dioxide, damp air 110

I

indicator, weather signs 8
inversion, warm atop cold 97
inversion, causes of 98
inverted "V", wave cyclone 49, 51

J

Jack Frost, conditions for frost 110

jet stream 69

K

kill, winter storm, wind chill 61

L

LA, fair and foul winds 73
lakes, cycle of water 22
lands, temperature retention 37
landscape, atmospheric, wind flow 16
latitudes,
    middle, air flow east to west 32
    middle, atmospheric events of 7
layers, stable air 97
leaves, estimating wind speed 67
leeward, side of a mountain 22
lift, lifting fog 109
lifting forces, outlined 20, 21
lightning, thunder storm 86
lightning strikes, level of 87
lines of equal pressure 16
locks of hair, cirrus clouds 79
love spark, Coriolis force 24
low clouds, types of 81

Words from rhymes appear in *italics* 127

low pressure,
    direction of flow around a low 29
    wind flow from high to low 16
lucky, guess, weather signs 8
lunch outdoors, morning dew 108

M

Maine, fair and foul winds 73
map, contour, wind flow 16
Miami, fair and foul winds 73
middle clouds, types of 81
middle latitudes,
    air flow west to east 32
    atmospheric events of 7
migration, seasonal, of the westerlies 37
moisture, cloud formation 18
monkeys, thunder storm 86
moon,
    estimating dew point 106
    ring around 114
    sharp, dull 115
morn, dewless 107
morning dew, skies of blue 109

morsels, sky 1
mountain, looks near before a storm 113
mountain's sides, lifting force 20
mud, friction and Coriolis 42
murky air, high dew point 106
music, clouds dance 3

N

nerves, ache of the weather 93
nest, sunset 116
newspapers, estimating wind speed 67
nitrogen, dry air 109, 110
north, passage of storm 60
northeast, wind from 76
northern hemisphere,
    air flow west to east 32
    atmospheric events of 7
northwest,
    cold quadrant 75
    fair-weather clouds 85
    wind from 94
nose, locating storm's center 51

O

o'rhead, barometer 91
observation, weather, general 5

occluded front, characteristics 58
oceans,
    cycle of water 22
    temperature retention 37
October, location of the westerlies 35
orange coat, Florida frost 10

P

palms, to depict a wave cyclone 47
particles, dry dust, sunset 117
percussion, rain plays 3
picnic, cloud bottoms 84
pinkies, to depict a wave cyclone 47
polar front,
    formation 49
    storm 51
pool, unseasonable summer temperatures 99
pouch, rain clouds 83
prefixes, cloud height 80
pressure,
    lines of equal pressure 16
    rising air 15
pressure's high, fair weather 92, 93
profusion, rain, occluded front 58
push and shove, unstable air 98

puzzle, weather, challenge to understand 3

Q

quadrant,
    northwest, cold 75
    northwest, fair-weather clouds 85
    storm quadrant characteristics 59
quilt,
    fog formation 107
    warm front 54

R

radiation fog, described 107
ragged feet, cloud bottoms, rain 84
rain,
    areas of a wave cyclone 55
    cold front 57
    occluded front 59
rainbow, morning, afternoon 118
rains, of spring and fall 32
ramp, warm front 53
rank and file, fair-weather clouds 85
red sky, sunset 117
refrain, cloud tops 84
ridge, upper-air 69-72
right, Coriolis force 25
ring, around moon and sun 114
rising, warm air 12, 13

river bed, upper-air westerlies 32
Rover, ache of the weather 93

S

screw, loosen, low pressure 29
sea, of air 2
seasons, starting dates 36, 37
Seattle, fair and foul winds 73
seesaw, winter storm 61
sequence, historical 6
shelter, rain clouds 83
shifts, wind
    causes of 45
    taking readings 74
shoulder, locating storm's center 51
showers, April, October 35
sick leave, dew, fair weather 108
sign, weather indicators 8
silvery moon, fair-weather clouds 87
sketch, high-level clouds, reliability 88
sky, a stage 1
sky, red sunset 117
slash, seasons, westerlies 35
sled, wind flow 16
slope, lifting force, air mass 20

smell, stormy weather 112
smell, general observation 6
smooth air, wind from southwest 76
snow, winter storm 61
soap, convergence, rising air 20
soul, sea of air 2
south, passage of storm 59
southeast,
    wind from 94
    wind, smoother 76
southwest, wind, smooth 76
space, winter temperatures 99
speed,
    weather-system 44
    westerly flow 32
spill it, rain 90
spinach, cloud bottoms, rain 84
spiraling, wind flow 44
spot, where hot meets cold 96
spots, of blue, breaking of the storm 88
spring, starting date 36
squeegee, dew, fair weather 108
stability, defined 97
stage, sky 1
star, twinkling 116
stars,
    bright, pale 115
    estimating the dew point 106

Words from rhymes appear in *italics*

stirring mechanism, global air flow 16
stool, steady rain, north quadrant 59
storm,
    breaking of 88
    direction of movement 71
    divergence aloft 71
    locating center 51
    passage to north 60
    passage to south 59
    polar front 51
    quadrant characteristics 59
    straight-on approach 75
storm belt, upper-air westerlies 35
story, every wind 65
straight, stratus clouds 79
stratocumulus clouds, described 81
stratus clouds, described 80, 81
straw, rising warm air 13
stream bed, upper-air westerlies 31
street, cloud street 85
stroll, fair-weather clouds 87
style, fair-weather clouds 85
subsidence, upper-air ridge 72
sultry morning, thunder 109

summer,
    starting date 36
    temperatures 99
sun,
    clouds thick enough to hide 83
    ring around 114
    wind from northwest 94
sunniest, rising barometer 94
sunset, red sky 117
sure enough, long rain 77
swirling force, Coriolis force 28
symmetry, weather system 45, 46

T

table, sky 1
tails, height of cirrus clouds 82
temperature,
    cause of winds 11
    frontal passage 57
temperature,
    contrasts 96
    dew point 103
    land, oceans 37
    seasonal changes 37
    unseasonable 99
ten, location of westerlies 34
thaw, winter storm 61
thick, enough to hide the sun 83
thumbs, to depict a wave cyclone 47

Words from rhymes appear in italics   131

thunder storms, path of 86
tides, sea of air 2
tilt, of the earth's axis 33
tomato, frost 110
tops, cloud tops, rain 84
towers, rain clouds 84
town, unpleasant weather 95
trace, unseasonable winter temperatures 99
track,
    of a storm 71
    of the westerlies 35
traditions, historical sequence 6
trails, contrails 89
trees, estimating wind speed 67
trend, predictors of warming 75
Tropic of Cancer, summer start, longest day 36
Tropic of Capricorn, winter start, shortest day 36
trough, upper-air westerlies 69
true, dew point 103
tune, wind sings 3
turrets, rain clouds 84
TV, watch the weather 3
twinkle, stars 116

U

unstable air, wind from northwest 76

updrafts, wind from northwest 76
upstream, westerly air flow 32

V

V, inverted "V", wave cyclone 51
vacuum, rising air 15
veering, wind 76

W

walks a lot, cool air sinks 12
wall, vertical, cold front 56
warm front,
    characteristics 52
    formation, wave cyclone 51
warm sector, of winter storm 62
water, cycle of 22
wave, polar front 49
wave cyclone,
    formation 49, 51
    top-view diagram 55
    typical 48
    winter 61
waves, sea of air 2
weather, average, climate 40
weather, air mass 38
wedge, cold front 57
weight,
    of dry air 109
    of damp air 110

westerlies, effect of location 32
westerlies, seasonal passage 35
westerly currents, speed of flow 32
wheat, wet spring 10
whistle, sound, loud before storm 113
wicked spell, wind stories 65
wind,
    caused by temperature differences 11
    frontal passage 57
wind,
    circular, caused by friction 42
    cycle of 16, 22
    direction of spiral 29
    from east, rain 94
    from northeast, steadier 76
    from northwest, fair 73
    from northwest, erratic 76
    from northwest, cold 94
    from southeast, rain 73
    from southeast, smoother 76
    from southwest, smooth 76
    general observation 6
    gusty, cold front 57
    judging speed by clouds 68
    local obstructions 67
    no shifts 75
    shifts, taking readings 74
    speed indicators 67
    spiraling in westerly stream 44
    affected by time of day 66
    upper-air, from north 72
    upper-air, from south 71
    veering 76
    which winds to trust 66
wind chill,
    described 100
    estimating 101
windows, sticking, humidity 105
windward, side of a mountain 22
winter,
    starting date 36
    unseasonable temperatures 99
    wave cyclone 61
witness, where hot meets cold 96
wits, watch the weather 3
wring, cycle of water 22

Y

yearning, Coriolis force 25